Hugo Hens

Performance Based Building Design 2

From Timber-framed Construction to Partition Walls

Other books from Hugo S. L. C. Hens at Ernst & Sohn:

Hugo S. L. C. Hens
Applied Building Physics
Boundary Conditions, Building Performance and Material Properties

2010 updated edition 2012.
322 pages. 101 fig. 20 tab.
Softcover.
€ 59,–
ISBN: 978-3-433-02962-6

 book

Hugo S. L. C. Hens
**Building Physics –
Heat, Air and Moisture**
Fundamentals and Engineering Methods with Examples and Exercises

329 pages. 142 fig. 28 tab.
Softcover.
€ 59,–
ISBN: 978-3-433-03027-1

 book

Hugo S. L. C. Hens
**Performance Based
Building Design 1**
From Below Grade Construction to Cavity Walls

2012.
276 pages. 172 fig.
Softcover.
€ 59,–
ISBN: 978-3-433-03022-6

 book

Set-Price: € 99,–
**Building Physics and
Applied Building Physics**
ISBN:
978-3-433-03031-1

Set-Price: € 99,–
**Performance Based
Building Design 1 and 2**
ISBN:
978-3-433-03024-0

Information and Order

www.ernst-und-sohn.de

Use online and Download

www.onlinelibrary.wiley.com

Hugo Hens

Performance Based Building Design 2

From Timber-framed Construction to Partition Walls

Professor Hugo S. L. C. Hens
University of Leuven (KULeuven)
Department of Civil Engineering
Building Physics
Kasteelpark Arenberg 40
3001 Leuven
Belgium

Cover: Low Energy Brick Building (Students' Residence), KULeuven, Belgium

Photo: Hugo Hens

Library of Congress Card No.:
applied for

British Library Cataloguing-in-Publication Data
A catalogue record for this book is available from the British Library.

**Bibliographic information published by
the Deutsche Nationalbibliothek**
The Deutsche Nationalbibliothek lists this publication in the Deutsche Nationalbibliografie; detailed bibliographic data are available on the Internet at http://dnb.d-nb.de.

© 2013 Wilhelm Ernst & Sohn, Verlag für Architektur und technische Wissenschaften GmbH & Co. KG, Rotherstr. 21, 10245 Berlin, Germany

All rights reserved (including those of translation into other languages). No part of this book may be reproduced in any form – by photoprinting, microfilm, or any other means – nor transmitted or translated into a machine language without written permission from the publishers. Registered names, trademarks, etc. used in this book, even when not specifically marked as such, are not to be considered unprotected by law.

Coverdesign: Sophie Bleifuß, Berlin, Germany
Typesetting: Manuela Treindl, Fürth, Germany
Printing and Binding: betz-Druck GmbH, Darmstadt, Germany

Printed in the Federal Republic of Germany.
Printed on acid-free paper.

Print ISBN: 978-3-433-03023-3
ePDF ISBN: 978-3-433-60248-5
ePub ISBN: 978-3-433-60249-2
mobi ISBN: 978-3-433-60250-8
oBook ISBN: 978-3-433-60251-5

To my wife, children and grandchildren

*In remembrance of Professor A. de Grave
who introduced building physics as a new discipline
at the University of Leuven, KULeuven, Belgium, in 1952*

Preface

Overview

Just like building physics, performance based building design was hardly an issue before the energy crises of the 1970s. With the need to upgrade energy efficiency, the interest in overall building performance grew. The volume on applied building physics discussed a performance rationale and performance requirements at the building and building enclosure level, with emphasis on heat, air, moisture checks. As in the third volume, volume four continues this rationale for structural aspects, acoustics, fire safety, maintenance and buildability. And as with volume three, it is the result of thirty-eight years of teaching architectural, building and civil engineers, coupled to more than forty years of experience in research and consultancy. Where and when needed, input and literature from around the world has been used, with a list of references and literature at the end of each chapter.

The book can be used by undergraduates and graduates in architectural and building engineering and also building engineers who want to refresh their knowledge may also benefit. The level of discussion assumes a sound knowledge of building physics, along with a background in structural engineering, building materials and building construction.

Acknowledgements

A book of this magnitude reflects the work of many, not only the author. Therefore, first of all, we want to thank our thousands of students. They gave us the opportunity to test the content and helped in upgrading by the corrections they proposed and the experience they offered in learning what parts should be better explained.

This is a text that has been written standing on the shoulders of those who came before us. Although we started our career as a structural engineer, our predecessor, Professor Antoine de Grave planted the seeds from which our interest in building physics, building services and performance based building design slowly grew. The late Bob Vos of TNO, the Netherlands, and Helmut Künzel of the Fraunhofer Institut für Bauphysik, Germany, showed the importance of experimental work and field testing for understanding building performance, while Lars Erik Nevander of Lund University, Sweden, taught that application does not always ask extended modeling, mainly because reality in building construction is much more complex than any simulation is.

During the four decades at the Laboratory of Building Physics, several researchers and Ph.D.-students got involved. I am very grateful to Gerrit Vermeir, Staf Roels, Dirk Saelens and Hans Janssen who became colleagues at the university; to Jan Carmeliet, now professor at the ETH-Zürich; Piet Standaert, a principal at Physibel Engineering; Jan Lecompte, at Bekaert NV; Filip Descamps, a principal at Daidalos Engineering and part-time professor at the Free University Brussels (VUB); Arnold Janssens, professor at the University of Ghent (UG); Rongjin Zheng, associate professor at Zhejiang University, China, and Bert Blocken, professor at the Technical University Eindhoven (TU/e), who all contributed by their work. The experiences gained by working as a structural engineer and building site supervisor at the start of my career, as building assessor over the years, as researcher and operating agent of four Annexes of the IEA, Executive Committee on Energy Conservation in Buildings and Community Systems, forced me to rethink the engineering based performance approach each time again. The many ideas I exchanged with Kumar Kumaran, Paul Fazio, Bill Brown,

William B. Rose, Joe Lstiburek and Anton Ten Wolde in Canada and the USA were also of great help. A number of reviewers took time to examine the book. Although we do not know their names, we also thank them here.

Finally, I thank my family, my wife Lieve, who managed living together with a busy engineering professor, my three children who had to live with that busy father and my many grandchildren who do not know their grandfather is still busy.

Leuven, June 2012						*Hugo S. L. C. Hens*

Contents

	Preface	VII
0	**Introduction**	1
0.1	Subject of the book	1
0.2	Units and symbols	1
0.3	References and literature	5
1	**Timber-framed construction**	7
1.1	In general	7
1.2	Performance evaluation	9
1.2.1	Structural integrity	9
1.2.2	Building physics: heat, air, moisture	11
1.2.2.1	Air tightness	11
1.2.2.2	Thermal transmittance	13
1.2.2.3	Transient response	16
1.2.2.4	Moisture tolerance	17
1.2.2.5	Thermal bridges	23
1.2.3	Building physics: acoustics	24
1.2.4	Durability	24
1.2.5	Fire safety	25
1.2.6	Maintenance	25
1.3	Design and execution	25
1.3.1	Above grade	25
1.3.2	Frame	25
1.3.3	Thermal insulation	26
1.3.4	Air and vapour retarder	26
1.3.5	Building paper	26
1.3.6	Variants	27
1.4	References and literature	27
2	**Sheet-metal outer wall systems**	31
2.1	In general	31
2.2	Performance evaluation	31
2.2.1	Structural integrity	31
2.2.2	Building physics: heat, air, moisture	31
2.2.2.1	Air tightness	32
2.2.2.2	Thermal transmittance	32
2.2.2.3	Transient response	33
2.2.2.4	Moisture tolerance	33
2.2.2.5	Thermal bridges	35

2.2.3	Building physics: acoustics	35
2.2.4	Durability	36
2.2.5	Fire safety	36
2.2.6	Maintenance	36
2.3	Design and execution	36
2.4	References and literature	37
3	**New developments**	**39**
3.1	Transparent insulation	39
3.1.1	In general	39
3.1.2	Performance evaluation	39
3.1.2.1	Building physics: heat, air, moisture	39
3.1.2.2	Durability	44
3.2	Multiple skin and photovoltaic outer walls	44
3.3	References and literature	44
4	**Roofs: requirements**	**45**
4.1	In general	45
4.2	Performance evaluation	45
4.2.1	Structural integrity	45
4.2.2	Building physics: heat, air, moisture	45
4.2.2.1	Air tightness	45
4.2.2.2	Thermal transmittance	45
4.2.2.3	Transient response	46
4.2.2.4	Moisture tolerance	47
4.2.2.5	Thermal bridges	48
4.2.3	Building physics: acoustics	48
4.2.4	Durability	48
4.2.5	Fire safety	49
4.2.6	Maintenance and economy	49
4.3	References and literature	49
5	**Low-sloped roofs**	**51**
5.1	Typologies	51
5.2	Roofing membranes	53
5.2.1	Build-up, multi-ply roofing	53
5.2.1.1	Number of layers	53
5.2.1.2	Execution	54
5.2.1.3	Bonding	54
5.2.1.4	Protection	55
5.2.1.5	Combination with the type of substrate	55
5.2.2	Build up polymer roofing	56
5.2.3	Problems with roofing	56
5.2.3.1	Pimples	56

5.2.3.2	Alligator skin	57
5.2.3.3	Cracking	57
5.2.3.4	Blistering	57
5.3	Compact low-sloped roofs	58
5.3.1	Assemblies	58
5.3.1.1	Heavy deck	58
5.3.1.2	Semi-heavy deck	59
5.3.1.3	Light-weight deck	59
5.3.1.4	Conclusion	59
5.3.2	Performance evaluation	60
5.3.2.1	Structural integrity	60
5.3.2.2	Building physics: heat, air, moisture	62
5.3.2.3	Building physics: acoustics	81
5.3.2.4	Fire safety	81
5.3.2.5	Maintenance	81
5.3.3	Design and execution	81
5.3.3.1	Assembly	81
5.3.3.2	Details	82
5.3.3.3	Special low-sloped roof uses	84
5.4	Protected membrane roofs	85
5.4.1	In general	85
5.4.2	Performance evaluation	86
5.4.2.1	Thermal transmittance	86
5.4.2.2	Moisture tolerance	90
5.4.2.3	Other performances	92
5.4.3	Design and execution	92
5.4.3.1	Roofing membrane	92
5.4.3.2	Details	93
5.4.3.3	Globally	93
5.5	References and literature	94
6	**Pitched roofs**	**97**
6.1	Classification	97
6.1.1	Type of form	97
6.1.1.1	Simple roofs	97
6.1.1.2	Composite roofs	98
6.1.2	Type of supporting structure	98
6.1.2.1	Purlins	98
6.1.2.2	Rafters	98
6.1.2.3	Other	99
6.1.3	Type of cover	99
6.2	Roof covers in detail	99
6.2.1	Tiles	99
6.2.1.1	Ceramic	99
6.2.1.2	Concrete	101
6.2.1.3	Metallic	102

6.2.2	Slates		102
6.2.2.1	Quarry		102
6.2.2.2	Fibre cement		103
6.2.2.3	Timber slates and plain tiles		103
6.2.3	Corrugated sheets		104
6.2.4	Shingles		104
6.3	Basic assemblies		105
6.4	Performance evaluation		107
6.4.1	Structural integrity		107
6.4.2	Building physics: heat, air, moisture		108
6.4.2.1	Air tightness		108
6.4.2.2	Thermal transmittance		114
6.4.2.3	Transient response		123
6.4.2.4	Moisture tolerance		123
6.4.2.5	Thermal bridges		139
6.4.3	Building physics: acoustics		141
6.4.4	Durability		141
6.4.5	Fire safety		142
6.4.6	Maintenance		142
6.5	Design and execution		143
6.5.1	Roof assemblies		143
6.5.1.1	Attic as storage and buffer space		143
6.5.1.2	Inhabited attic		143
6.5.2	Roof details		147
6.6	References and literature		149
7	**Sheet-metal roofs**		**153**
7.1	In general		153
7.2	Metal roof cover		154
7.2.1	Overview		154
7.2.2	Lead		156
7.2.3	Copper and brass		157
7.2.4	Zinc		158
7.2.5	Aluminium		158
7.3	Performance evaluation		159
7.3.1	Moisture tolerance		159
7.3.1.1	In general		159
7.3.1.2	Lack of air tightness		160
7.3.1.3	Under cooling		161
7.3.1.4	Two practice and six test building cases		163
7.3.2	Thermal bridges		168
7.3.3	Durability		169
7.4	Design and execution		170
7.5	References and literature		170

Contents   XIII

8	**Windows, outer doors and glass façades**	173
8.1	In general	173
8.2	Glass	173
8.2.1	In general	173
8.2.2	Performance evaluation	174
8.2.2.1	Structural safety	174
8.2.2.2	Building physics: heat, air, moisture	177
8.2.2.3	Building physics: acoustics	194
8.2.2.4	Building physics: light	196
8.2.2.5	Durability	197
8.2.2.6	Fire safety	198
8.2.2.7	Break-in safety	199
8.2.3	Technology	199
8.3	Windows and doors	200
8.3.1	In general	200
8.3.2	Performance evaluation	204
8.3.2.1	Structural safety	204
8.3.2.2	Building physics: heat, air, moisture	204
8.3.2.3	Building physics: acoustics	210
8.3.2.4	Durability	211
8.3.2.5	Fire safety	211
8.3.2.6	Maintenance and safety at use	211
8.3.3	Technology	212
8.3.3.1	Timber, vinyl	212
8.3.3.2	Aluminium, steel	212
8.3.3.3	Window blinds and roller shutters	212
8.3.3.4	Trickle vents	213
8.4	Glass façades	214
8.4.1	In general	214
8.4.2	Window fronts	214
8.4.3	Curtain walls	215
8.4.4	Structural and suspended glazing	217
8.4.5	Double skin façades	218
8.4.5.1	In general	218
8.4.5.2	Building physics: heat, air, moisture	218
8.4.5.3	Building physics: acoustics	225
8.4.5.4	Fire safety	226
8.4.5.5	Building level: energy efficiency	226
8.4.6	Photovoltaic façades (PV)	226
8.5	References and literature	226
9	**Balconies, shafts, chimneys and stairs**	231
9.1	In general	231
9.2	Balconies	231
9.2.1	In general	231

9.2.2	Performance evaluation	231
9.2.2.1	Structural integrity	231
9.2.2.2	Building physics: heat, air, moisture	231
9.2.2.3	Durability	233
9.2.2.4	Fire safety	234
9.2.2.5	User's safety	234
9.2.3	Design and execution	234
9.2.3.1	Thermal cuts	234
9.2.3.2	Water proofing and drainage	234
9.2.3.3	Floor finish	235
9.2.3.4	Hand rails	235
9.3	Shafts	235
9.3.1	In general	235
9.3.2	Performance evaluation	235
9.3.2.1	Structural integrity	236
9.3.2.2	Building physics	236
9.3.2.3	Fire safety	236
9.3.3	Design and execution	236
9.4	Chimneys	237
9.4.1	In general	237
9.4.2	Design considerations	237
9.4.3	Design and execution	239
9.4.3.1	Stoves	239
9.4.3.2	Boilers	239
9.5	Stairs	240
9.5.1	In general	240
9.5.2	Performance evaluation	242
9.5.2.1	Structural integrity	242
9.5.2.2	Building physics: acoustics	243
9.5.2.3	Fire safety	244
9.5.2.4	User's safety	244
9.5.3	Design and execution	244
9.6	References and literature	244
10	**Partitions; wall, floor and ceiling finishes; inside carpentry**	**245**
10.1	Overview	245
10.2	Partition walls	245
10.2.1	In general	245
10.2.2	Performance evaluation	246
10.2.2.1	Structural integrity	246
10.2.2.2	Building physics: heat, air, moisture	246
10.2.2.3	Building physics: acoustics	248
10.2.2.4	Durability	249
10.2.2.5	Fire safety	249
10.2.3	Design and execution	249

10.3	Building services	250
10.4	Wall finishes	250
10.4.1	In general	250
10.4.2	Performance evaluation	250
10.4.2.1	Structural integrity	250
10.4.2.2	Building physics: heat, air, moisture	251
10.4.2.3	Building physics: acoustics	251
10.4.2.4	Fire safety	251
10.4.3	Design and execution	251
10.4.3.1	Wet plasters	251
10.4.3.2	Gypsum board	252
10.5	Floor finishes	252
10.5.1	In general	252
10.5.2	Performance evaluation	253
10.5.2.1	Structural integrity	253
10.5.2.2	Building physics: heat, air, moisture	254
10.5.2.3	Building physics: acoustics	255
10.5.2.4	Durability	256
10.5.3	Design and execution	256
10.6	Ceiling finishes	257
10.6.1	In general	257
10.6.2	Performance evaluation	258
10.6.2.1	Structural integrity	258
10.6.2.2	Building physics: heat, air, moisture	258
10.6.2.3	Building physics: acoustics	259
10.6.2.4	Fire safety	259
10.6.3	Design and execution	259
10.7	Inside carpentry	260
10.7.1	In general	260
10.7.2	Performance evaluation	260
10.7.2.1	Structural integrity	260
10.7.2.2	Building physics: heat, air, moisture	261
10.7.2.3	Building physics: acoustics	261
10.7.2.4	Fire safety	262
10.7.3	Design and execution	263
10.8	References and literature	263
11	**Risk analysis**	**265**
11.1	In general	265
11.2	Risk definition	265
11.3	Performing a risk analysis	266
11.3.1	Identification and probability of deficiencies	266
11.3.2	Severity of the consequences	266
11.3.3	Proposals to limit risks	266

11.4	Example of risk analysis: cavity walls	267
11.4.1	Generalities	267
11.4.2	Deficiencies encountered	267
11.4.3	Probabilities	269
11.4.4	Severity of the consequences	269
11.4.5	Risk?	271
11.4.5.1	Mould	271
11.4.5.2	Rain penetration	272
11.4.5.3	Others	273
11.4.6	Evaluation	273
11.4.7	Upgrade proposals	274
11.5	References and literature	274

0 Introduction

0.1 Subject of the book

This is the second part of the third volume in a series of books on building physics, applied building physics and performance based building design:

- Building Physics: Heat, Air and Moisture
- Applied Building Physics: Boundary Conditions, Building Performance and Material Properties
- Performance Based Building Design 1
- **Performance Based Building Design 2**

Performance Based Building Design 2 continues the application of the performance based engineering rationale, discussed in 'Applied Building Physics: Boundary Conditions, Building Performance and Material Properties' to the design and construction of building assemblies. In order to do that, the text considers the performance requirements presumed or imposed, their prediction during the design stage and the technology needed for realization.

Performance Based Building Design 1 ended with massive outer walls. Performance Based Building Design 2 begins with lightweight building and outer wall systems: timber-framed and metal-based. Then low-sloped, pitched, and metal roofs follow to finish the enclosure-related subjects with glazed surfaces and windows. Attention then turns to balconies, chimneys, shafts, staircases, inside partitions, and finishes. The volume closes with a chapter on risk analysis. Of course, for principals acceptable risk is an important issue. As in Performance Based Building Design 1, the impact of performance requirements on design and execution is highlighted. For decades, the Laboratory of Building Physics at the KULeuven not only tested highly insulated massive façade assemblies, but also lightweight façade assemblies and roofs. The results are used in the discussions.

0.2 Units and symbols

The book uses the SI-system (internationally mandatory since 1977). Base units are the meter (m), the kilogram (kg), the second (s), the Kelvin (K), the ampere (A) and the candela. Derived units of importance are:

Unit of force: Newton (N); $1 \text{ N} = 1 \text{ kg} \cdot \text{m} \cdot \text{s}^{-2}$
Unit of pressure: Pascal (Pa); $1 \text{ Pa} = 1 \text{ N/m}^2 = 1 \text{ kg} \cdot \text{m}^{-1} \cdot \text{s}^{-2}$
Unit of energy: Joule   (J); $1 \text{ J} = 1 \text{ N} \cdot \text{m} = 1 \text{ kg} \cdot \text{m}^2 \cdot \text{s}^{-2}$
Unit of power: Watt   (W); $1 \text{ W} = 1 \text{ J} \cdot \text{s}^{-1} = 1 \text{ kg} \cdot \text{m}^2 \cdot \text{s}^{-3}$

For the symbols, the ISO-standards (International Standardization Organization) are followed. If a quantity is not included in these standards, the CIB-W40 recommendations (International Council for Building Research, Studies, and Documentation, Working Group 'Heat and Moisture Transfer in Buildings') and the list edited by Annex 24 of the IEA, ECBCS (International Energy Agency, Executive Committee on Energy Conservation in Buildings and Community Systems) are applied.

Table 0.1. List with symbols and quantities.

Symbol	Meaning	Units
a	Acceleration	m/s^2
a	Thermal diffusivity	m^2/s
b	Thermal effusivity	W/(m$^2 \cdot$ K $\cdot$ s$^{0.5}$)
c	Specific heat capacity	J/(kg $\cdot$ K)
c	Concentration	kg/m^3, g/m^3
e	Emissivity	–
f	Specific free energy	J/kg
	Temperature ratio	–
g	Specific free enthalpy	J/kg
g	Acceleration by gravity	m/s^2
g	Mass flow rate, mass flux	kg/(m$^2 \cdot$ s)
h	Height	m
h	Specific enthalpy	J/kg
h	Surface film coefficient for heat transfer	W/(m$^2 \cdot$ K)
k	Mass related permeability (mass may be moisture, air, salt …)	s
l	Length	m
l	Specific enthalpy of evaporation or melting	J/kg
m	Mass	kg
n	Ventilation rate	s^{-1}, h^{-1}
p	Partial pressure	Pa
q	Heat flow rate, heat flux	W/m^2
r	Radius	m
s	Specific entropy	J/(kg $\cdot$ K)
t	Time	s
u	Specific latent energy	J/kg
v	Velocity	m/s
w	Moisture content	kg/m^3
x, y, z	Cartesian co-ordinates	m
A	Water sorption coefficient	kg/(m$^2 \cdot$ s$^{0.5}$)
A	Area	m^2
B	Water penetration coefficient	m/s$^{0.5}$
D	Diffusion coefficient	m^2/s
D	Moisture diffusivity	m^2/s
E	Irradiation	W/m^2

0.2 Units and symbols

Table 0.1. (continued)

Symbol	Meaning	Units
F	Free energy	J
G	Free enthalpy	J
G	Mass flow (mass = vapour, water, air, salt)	kg/s
H	Enthalpy	J
I	Radiation intensity	J/rad
K	Thermal moisture diffusion coefficient	kg/(m·s·K)
K	Mass permeance	s/m
K	Force	N
L	Luminosity	W/m^2
M	Emittance	W/m^2
P	Power	W
P	Thermal permeance	W/(m^2·K)
P	Total pressure	Pa
Q	Heat	J
R	Thermal resistance	m^2·K/W
R	Gas constant	J/(kg·K)
S	Entropy, saturation degree	J/K, –
T	Absolute temperature	K
T	Period (of a vibration or a wave)	s, days, etc.
U	Latent energy	J
U	Thermal transmittance	W/(m^2·K)
V	Volume	m^3
W	Air resistance	m/s
X	Moisture ratio	kg/kg
Z	Diffusion resistance	m/s
α	Thermal expansion coefficient	K^{-1}
α	Absorptivity	–
β	Surface film coefficient for diffusion	s/m
β	Volumetric thermal expansion coefficient	K^{-1}
η	Dynamic viscosity	N·s/m^2
θ	Temperature	°C
λ	Thermal conductivity	W/(m·K)
μ	Vapour resistance factor	–
ν	Kinematic viscosity	m^2/s

Symbol	Meaning	Units
ρ	Density	kg/m^3
ρ	Reflectivity	–
σ	Surface tension	N/m
τ	Transmissivity	–
ϕ	Relative humidity	–
α, ϕ, Θ	Angle	rad
ξ	Specific moisture capacity	kg/kg per unit of moisture potential
Ψ	Porosity	–
Ψ	Volumetric moisture ratio	m^3/m^3
Φ	Heat flow	W

Table 0.2. List with suffixes and notations.

Symbol	Meaning
Indices	
A	Air
c	Capillary, convection
e	Outside, outdoors
h	Hygroscopic
i	Inside, indoors
cr	Critical
CO_2, SO_2	Chemical symbol for gases
m	Moisture, maximal
r	Radiant, radiation
sat	Saturation
s	Surface, area, suction
rs	Resulting
v	Water vapour
w	Water
ϕ	Relative humidity
Notation	
[ ], bold	Matrix, array, value of a complex number
Dash	Vector (ex.: $\bar{a}$)

0.3 References and literature

[0.1] CIB-W40 (1975). Quantities, Symbols and Units for the description of heat and moisture transfer in Buildings: Conversion factors, IBBC-TNP, report No. BI-75-59/03.8.12, Rijswijk.

[0.2] ISO-BIN (1985). Standards series X02-101 – X023-113.

[0.3] Kumaran, K. (1996). *Task 3: Material Properties.* Final Report IEA EXCO ECBCS Annex 24. ACCO, Louvain, p. 135.

1 Timber-framed construction

1.1 In general

In the Low Countries on the North Sea, timber was the common construction material for rural and municipal dwellings until the $13^{th}-14^{th}$ century. Brick construction was an aristocrat's privilege. Many devastating town fires, the sociological fact that bricks stood for wealth and growing wood shortages slowly turned brick building into the new standard.

Timber construction still is the reference in many countries worldwide, like the US, Canada, Norway, Sweden, Finland, Russia, Japan and other countries rich in forests and often with a cold climate. There, the framed type has an important advantage compared to massive construction: it is easy to insulate, which is why even in northwest Europe timber-frame construction has regained popularity, now for passive houses. However, the disadvantages also deserve mentioning: hardly any thermal inertia, air tightness critical and less moisture tolerant than brick construction.

In timber framing, load- and non-bearing outer and partitions walls consist of a framework of timber studs and crossbeams, called plates. The outer wall frames are externally finished with structural sheathing. Where the studs bear all vertical loads and the outer wall ones have also to withstand the wind component, normal to the façade, the sheathing provides overall stiffness against horizontal loading. It also prevents buckling of the studs parallel to their lowest inertia radius. From the three common framing approaches – platform, balloon, post and beam – the platform type, composed of storey-high stud walls and timber floors is the most popular (Figure 1.1).

Construction looks as follows: once the foundations and foundation walls are ready, the ground floor is laid, in humid climates preferably a concrete deck, though in dry climates also timber joists with plywood or OSB (oriented strand board) deck apply, the crosscut end sides being closed with header plates. In such case, ripped half-width standard timber beams form the floor joists with struts at half-span excluding lateral buckling. Then one fixes the bottom plates, after which the studs are nailed and coupled with top plates. To stabilize the frame corners, doubling these is an option. After, a plywood, OSB or stiff insulation board (XPS) sheathing is nailed to the outer wall frames. The joists of the second floor, which are fixed at the top plates then follow. Header plates again close the crosscut end sides and plywood or OSB forms the running surface. The same cycle restarts for the second storey: bottom plate, studs, top plates, sheathing, floor joists, running surface, etc.

A timber framework or rafters, axis to axis at the same distance as the studs, shape the load-bearing roof structure with an external sheathing once more providing stiffness. Timber framing ends with wrapping up the outer walls with waterproof, wind tight building paper, stapled from bottom to top on the sheathing with the higher strips overlapping the lower ones. Platform framing lends itself to modular construction and prefabrication.

From inside to outside the outer wall assembly looks like (Figure 1.2): inside lining (gypsum board); (service cavity); air (always) and vapour (when necessary) retarder; bays between studs filled with insulation (mineral wool or cellulose); plywood, OSB or stiff insulation board sheathing; building paper; outside finish (timber siding, brick veneer, EIFS, etc).

Aside from timber framing, also metal framed construction exists, with metal studs and plates replacing the timber ones.

Performance Based Building Design 2. From Timber-framed Construction to Partition Walls.
First Edition. Hugo Hens.
© 2013 Ernst & Sohn GmbH & Co. KG. Published 2013 by Ernst & Sohn GmbH & Co. KG.

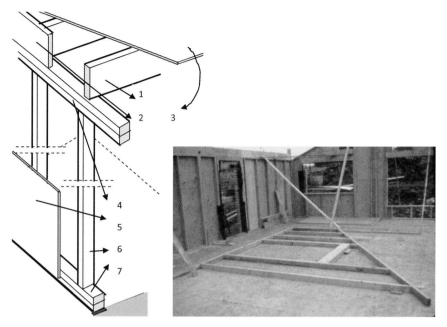

Figure 1.1. Platform type (1: joists, 2: header plate, 3: running surface, 4: top plates, 5: sheathing, 6: studs, 7: bottom plates).

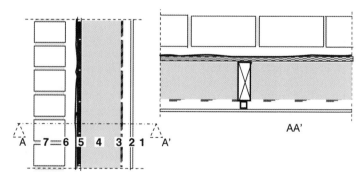

Figure 1.2. Timber-framed outer wall, reference assembly (1: inside lining, 2: service cavity, 3: air and vapour retarder, 4: thermal insulation; 5: sheathing, 6: building paper, 7: outside finish).

1.2 Performance evaluation

1.2.1 Structural integrity

Timber-framed buildings are so lightweight that anchoring in the foundation walls is necessary to prevent displacement under extreme wind load (Figure 1.3).

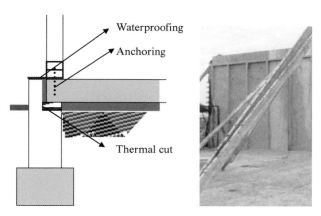

Figure 1.3. Timber-framed construction, anchoring in the foundation walls.

Wind loading and buckling of the outer and partition wall studs demands proper attention. The sheathing or inside finishes block it in the lowest moment of inertia direction. The direction normal to the walls needs a control. Table 1.1 gives the buckling factors vertical loads have to be multiplied by, as a function of the stud's slenderness (i):

$$i = \frac{L}{\sqrt{\frac{I}{A}}} \tag{1.1}$$

with L the effective stud span (in timber framed construction equal to the distance between bottom and top plates), I the moment of inertia around the neutral axis of the combination stud/sheathing (if shear-stiff coupled) and A total active cross section.

If this product gives stresses in the timber beyond acceptable, or, if for a given span the stud's radius of inertia is too low, then two options are left: diminishing the centre-to-centre distance between studs or using deeper ones. The first is disadvantageous in terms of whole wall thermal transmittance whereas the second allows larger insulation thicknesses, thus, a lower whole wall thermal transmittance.

Table 1.2 summarizes the mechanical properties of softwood and plywood. For the stiffness against horizontal loads, the same rules as for massive construction hold: the floors as rigid horizontal decks, at least 3 sheathed or wind-braced walls whose centre planes do not cross in one point, the stiff walls preferentially distributed in a way the resulting wind load vector crosses their stiffness centre.

Table 1.1. Buckling factors (slenderness vertically in steps of 10, horizontally in steps of 1).

Slenderness	0	1	2	3	4	5	6	7	8	9
0	1	1	1.01	1.01	1.02	1.02	1.02	1.03	1.03	1.04
10	1.04	1.04	1.05	1.05	1.06	1.06	1.06	1.07	1.07	1.08
20	1.08	1.09	1.09	1.10	1.11	1.11	1.12	1.13	1.13	1.14
30	1.15	1.16	1.17	1.18	1.19	1.20	1.21	1.22	1.24	1.25
40	1.26	1.27	1.29	1.30	1.32	1.33	1.35	1.36	1.38	1.40
50	1.42	1.44	1.46	1.48	1.50	1.52	1.54	1.56	1.58	1.60
60	1.62	1.64	1.67	1.69	1.72	1.74	1.77	1.80	1.82	1.85
70	1.88	1.91	1.94	1.97	2.00	2.03	2.06	2.10	2.13	2.16
80	2.20	2.23	2.27	2.31	2.35	2.38	2.42	2.46	2.50	2.54
90	2.58	2.62	2.66	2.70	2.74	2.78	2.82	2.87	2.91	2.95
100	3.00	3.06	3.12	3.18	3.24	3.31	3.37	3.44	3.50	3.57
110	3.63	3.70	3.76	3.83	3.90	3.97	4.04	4.11	4.18	4.25
120	4.32	4.39	4.46	4.54	4.61	4.68	4.76	4.84	4.92	4.99
130	5.07	5.15	5.23	5.31	5.39	5.47	5.55	5.63	5.71	5.80
140	5.88	5.96	6.05	6.13	6.22	6.31	6.39	6.48	6.57	6.66
150	6.75	6.84	6.93	7.02	7.11	7.21	7.30	7.39	7.49	7.58
160	7.68	7.78	7.87	7.97	8.07	8.17	8.27	8.37	8.47	8.57
170	8.67	8.77	8.88	8.98	9.08	9.19	9.29	9.40	9.61	9.61
180	9.72	9.83	9.94	10.05	10.16	10.27	10.38	10.49	10.60	10.72
190	10.83	10.94	11.06	11.17	11.29	11.41	11.52	11.64	11.76	11.88
200	12.00	12.12	12.24	12.36	12.48	12.61	12.73	12.85	12.98	13.10
210	13.23	13.36	13.48	13.61	13.74	13.87	14.00	14.13	14.26	14.39
220	14.52	14.65	14.79	14.92	15.05	15.19	15.32	15.46	15.60	15.73
230	15.87	16.01	16.15	16.29	16.43	16.57	16.71	16.85	16.99	17.14
240	17.28	17.42	17.57	17.71	17.86	18.01	18.15	18.30	18.45	18.60

1.2 Performance evaluation

Table 1.2. Mechanical properties of softwood and plywood.

Property			Class 1	Class 2	Class 3	// fibres outer laminates	+ fibres outer laminates
			Softwood			Plywood	
Modulus of elasticity		MPa					
// fibres				11 000		7 000	
⊥ fibres				300		3 000	
Shear modulus		MPa		500			
Allowed stress							
Bending	// fibres	MPa	7	10	13		
	⊥ plywood	MPa				13	5
	// plywood	MPa				9	6
Tension	// fibres	MPa	0	8.5	10.5		
	// plywood	MPa				8	4
Compression	// fibres	MPa	6	8.5	11		
	⊥ fibres	MPa	2	2	2		
	⊥ plywood	MPa				3	3
	// plywood	MPa				8	4
Shear	// fibres	MPa	0.9	0.9	0.9		
	⊥ plywood	MPa				1.8	1.8
	// plywood	MPa				0.9	0.9

1.2.2 Building physics: heat, air, moisture

1.2.2.1 Air tightness

Air tightness of timber-framed envelopes is not taken for granted. The outside finish, the building paper, the sheathing, as well as the insulation, all are air-permeable. Contributing factors are, for the building paper, the overlaps between the strips, for the sheathing the joints between boards and for the thermal insulation the material itself and the gaps between insulation, studs and plates. It is the inside finish to guarantee air-tightness. Non-perforated gypsum board linings without cracks between boards have an air permeance of $(K_a) \approx 3.1 \cdot 10^{-5} \Delta P_a^{-0.19}$. For an air pressure difference of 10 Pa, that value limits air leakage to 0.43 m³/(m²·h). However, when sockets and others perforate the lining and cracks form between boards, this value may increase by a factor of 10, which is why inclusion of an additional air barrier deserves recommendation. In moderate and cold climates, one used a PE-foil, stapled against the timber frame, preferentially with a service cavity left between foil and inside lining. Recently, OSB with taped joints emerged as an alternative (Figure 1.4). But also with additional air barrier, perfect air-tightness is hard to realize. Even excellent workmanship did not result in tested air leakages below 3 dm³/(m²·h) at 1 Pa air pressure difference. In hot and humid climates, it is up to the outside finish to guarantee air-tightness.

Figure 1.4. Taped OSB as air barrier.

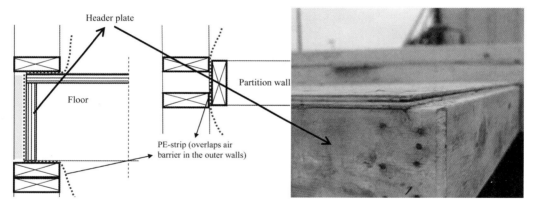

Figure 1.5. Timber-framed construction: caring for a continuous air barrier in the envelope.

Also three-dimensionally, a timber-framed construction offers a network of leaks. Via the junctions with the envelope, outside air may permeate partition walls, while conversely inside air can flow to the outside through the sockets in the partitions. At each floor level, air may flow façade to façade between the joists, a phenomenon causing unexpectedly high heat losses, quick ceiling soiling, and mould where the outside air enters. All this demands an envelope with continuous air barrier. Therefore, the following recommendations prevail: (1) include PE-strips at each floor between header plate and header insulation, (2) fix PE-strips in all junctions between outer and partition walls, (3) tape the overlaps to the air barrier (Figure 1.5).

Fully filling the space between sheathing and air barrier prevents air looping along the thermal insulation. A hotbox test on a two meters high timber framed wall insulated with 8 cm thick XPS-boards demonstrated that partial fills are critical. These are too stiff to link up perfectly with studs, plates, sheathing, and inside lining, creating leaks across and air layers at both sides of the insulation that way. At a temperature difference of 18.7 °C there was no uniform heat loss of 4.5 W/m^2 but large differences between the flow rates up and down the inside and outside surface were noted, see Table 1.3.

1.2 Performance evaluation

Table 1.3. Hot box test: heat flow rate across a timber-framed wall ($U_o = 0.24$ W/(m² · K)).

Height m	Heat flow rate W/m²	
	Outside surface	Inside surface
1.7	30.9	3.7
0.3	5.7	11.5

The reason is air looping, with cold air rising at the warm side of the insulation, warm air falling at the cold side, changeover from warm to cold on top of the insulation and changeover from cold to warm down the insulation. The data also suggest that thermal stack between hot and cold box activates outflow up, and inflow down the wall.

The building paper wrap should guarantee wind-tightness.

1.2.2.2 Thermal transmittance

The discussion relates to outer walls only. For roofs and floors, reference is made to the chapter on floors in Performance Based Building Design 1 and the chapters that follow on roofs. As always, the clear and whole wall thermal transmittances (U) differ, the last accounting for studs, top and bottom plates. In the case of an airtight outer wall, the series/parallel circuit of Figure 1.6 allows a fair guess of the whole wall thermal transmittance, as do also the following linear thermal transmittances (ψ):

Stud	Bottom plate	Top plates
$\psi = 0.017$ W/(m · K)	$\psi = 0.010$ W/(m · K)	$\psi = 0.023$ W/(m · K)

With mineral wool or cellulose as thermal insulation and a brick veneer as outside finish, the thicknesses of Table 1.4 give whole wall thermal transmittances of 0.4, 0.2 and 0.1 W/(m² · K) for 40 and 60 cm centred studs.

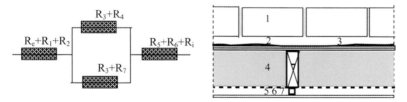

Figure 1.6. Timber framed wall as series/parallel circuit.

Table 1.4. Timber framed outer wall: insulation thicknesses (first number using ψ's, second according to series/parallel circuit).

U-value W/(m² · K)	Insulation thickness in cm			
	40 cm centred studs		60 cm centred studs	
	MW	Cellulose	MW	Cellulose
0.4	8/8	8/9	7/8	8/8
0.2	23/21	24/22	20/20	22/21
0.1	80//46	86/48	60/44	64/46

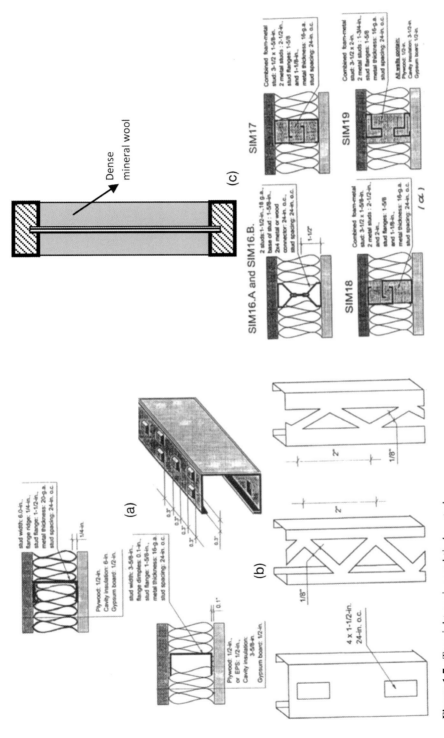

Figure 1.7. Top right engineered timber stud.
(a), (b), (c) are the steel shape studs Table 1.5 is based on.

1.2 Performance evaluation

For the values 0.4 and 0.2 W/(m² · K) both methods fit. Yet, for a value 0.1 W/(m² · K), the gap is manifest, showing that such low value demands a three-dimensional calculation. 0.2 W/(m² · K) gives wall thicknesses touching an acceptable 40 cm. Instead, 0.1 W/(m² · K) needs economically questionable thicknesses. The use of engineered studs and plates (Figure 1.7) gives some relief.

With metal frames, thermal bridging effects are more pronounced, with the Figure 1.6 circuit giving no reliable results anymore. Only measurement or three-dimensional calculations do. Take the first wall in Table 1.5. Its studs consist of cold-formed U-steel shapes with wall thickness 1.2 mm. Compared to the clear wall thermal resistance, the whole wall value drops by 38.2%, while timber studs limit that drop to 8.8%.

XPS as sheathing material, plus a smaller contact area between sheathing and steel studs or the use of perforated or thermally cut steel shapes gives the best results. The last bring the whole wall thermal transmittance in line with timber-framed walls.

In addition, the impact of workmanship when insulating the bays has been studied experimentally. Figure 1.8 shows some typical imperfections, while Table 1.6 lists their measured effect on the whole wall thermal transmittance. Increase peaks when air looping develops as is the case with narrowly cut insulation, creating 50 mm wide leaks at both studs.

Table 1.5. Clear and whole wall thermal resistance of the steel framed walls of Figure 1.7.

Assembly **61 cm centres**	R_0 m² · K/W	R Measured m² · K/W	R_1 Calculated m² · K/W	$\Delta R/R_0$ %
U-steel shapes 9.2 cm deep, plywood sheathing (Figure 1.7a)	2.25	1.39		38.2
U-steel shapes 9.2 cm deep, 2.5 cm XPS-sheathing	3.10	2.41		22.3
Steel shapes with met nipples 8.9 cm deep, plywood sheathing (Figure 1.7b)	2.25		1.54	31.6
Perforated steel shapes 9.2 cm deep, plywood sheathing (Figure 1.7c)	2.25		1.74	22.7
Perforated steel shapes 9.2 cm deep, 2.5 cm XPS-sheathing	2.81		2.42	14.0
Steel shapes with thermal cut, 8.9 cm deep, plywood sheathing (Figure 1.7d)	2.25		2.10	7.0

Table 1.6. Whole wall thermal transmittance in case of workmanship inaccuracies.

Timber studs 60 cm centre, 15 cm MW, **imperfections of Figure 1.8**	U_{meas} Measured W/(m² · K)	U Reference W/(m² · K)	$\Delta U/U$ %
None	0.230	0.230	0
Boards too strongly pressed against the studs	0.238	0.230	3.5
Insulation carelessly cut, wedge-shaped at studs	0.263	0.230	14.3
Insulation narrowly cut, 50 mm leak at one of the studs	0.246	0.230	7.0
Insulation narrowly cut, 50 mm leaks at both studs	0.350	0.230	50

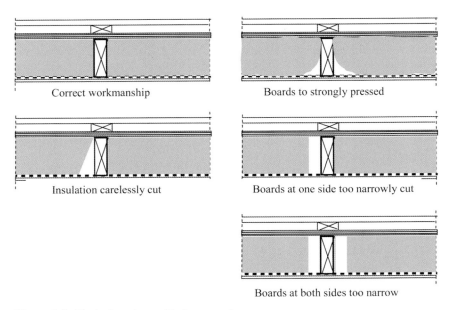

Figure 1.8. Typical workmanship inaccuracies.

1.2.2.3 Transient response

On a daily basis, timber-framed outer walls have an admittance way below 3.9 W/(m² · K) (for a surface film coefficient indoors of 7.8 W/(m² · K)), while the dynamic thermal resistance hardly differs from the steady state thermal resistance and temperature damping does not even approach a value 15. Better thermal insulation hardly changes things, see Table 1.7.

Table 1.7. Temperature damping, dynamic thermal resistance, and admittance (1-day period).

Wall, brick veneer as outside finish	Temperature damping + faze −, h		Dynamic thermal resistance + faze m² · K/W, h		Admittance + faze W/(m² · K), h	
4 cm MW, U_o = 0.47 W/(m² · K)	2.1	7.0	2.8	4.2	0.74	2.9
14 cm EPS, U_o = 0.21 W/(m² · K)	4.3	9.3	6.6	5.0	0.65	4.3

Through that, limited glass area, effective solar shading, and well-designed nighttime ventilation gain importance in moderate climates. Of course, an alternative is to combine a timber-framed envelope with heavy weight inside partitions and floors. To underline the difference, Figure 1.9 gives the fabric related room damping as function of window area for a room with a volume of 4 × 4 × 2.7 m³, a 4 × 2.7 m² timber-framed outer wall, clear wall thermal transmittance of 0.16 W/(m² · K), timber framed partition walls and joisted floors and, for the same room but now with brick partitions and concrete floors. With massive inside partitions and floors, damping increases by a factor of 4.

1.2 Performance evaluation

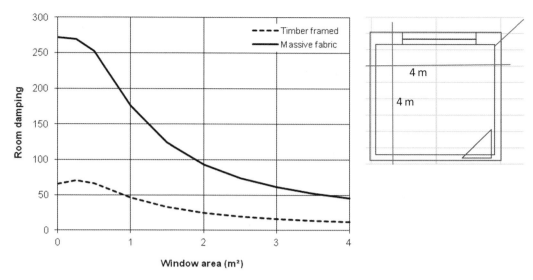

Figure 1.9. Fabric related room damping, integral timber framed versus outer wall only, in combination with massive partition walls and floors.

1.2.2.4 Moisture tolerance

Due to water sensitivity of the softwood used, timber-framed construction is inherently less moisture tolerant than massive construction. Above a moisture ratio of 20% kg/kg the risk to see mould colonizing the timber increases sharply whereas above 30% kg/kg fungal attack and bacterial rot become likely. To avoid problems the following requirements should be fulfilled:

1. Building moisture in studs, plates and joist must dry without damage
2. Once the construction is finished, rain should no longer seep in and humidify either the sheathing or the timber frame
3. Studs and plates should not suck water out of capillary porous materials they contact
4. Annually cumulating interstitial condensate is not allowed while a too high winter relative humidity lifting moisture ratio in the sheathing and frame beyond 20% kg/kg is excluded
5. Solar driven vapour flow giving moisture build-up in the insulation and moisture deposit against the air and vapour retarder or the inside lining should be avoided

Requirement 1

A vapour permeable outside finish facilitates fast drying of building moisture. Tests in the moderate, humid climate of Newfoundland, Canada, on eight walls proved building paper with low diffusion resistance is quite effective. All walls had a PE air and vapour retarder at their inside. Wall 1 and 2 were insulated with 14 cm mineral wool. Their frame was OSB sheathed and covered with building paper. Insulation in walls 3 to 6 was 8 cm mineral wool. For 3 and 4 the sheathing consisted of dense, 38 mm thick mineral wool boards covered with a vapour permeable spun-bonded foil. 5 and 6 had a 38 mm thick XPS sheathing, covered with the same spun-bonded foil. Wall 7 was insulated with 14 cm wet sprayed cellulose and

Table 1.8. Drying of timber framed walls (St John's, New Foundland).

Wall	R_1-value	µd sheathing + building paper	Building moisture	Moisture ratio after 1 year, % kg/kg			
				North		South	
	m² · K/W	m	% kg/kg	U	D	U	D
1.	4.1	4.3	26–30	21	31	20	20
2.	4.3	4.3	26–30	25	35	18	20
3.	3.8	0.01	26–30	12	15	12	16
4.	3.9	0.01	26–30	11	15	10	15
5.	3.9	5.9	26–30	23	29	18	25
6.	4.1	5.9	26–30	20	27	15	18
7.	3.9	4.3	26–30	45	76	68	118
8.	3.7	3.9	26–30	10	17	14	15

U = up, D = down

finished with an OSB sheathing. Wall 8 finally got 127 mm EPS as insulation, which was covered with a vapour permeable foil. All walls had humid studs and plates with a building moisture ratio from 26 to 30% kg/kg. Table 1.8 gives the measured moisture ratio after one-year exposure.

Walls facing south dried faster than north facing ones. After 1 year, the studs of walls 3 and 4, the one with vapour permeable finish, are driest, with a moisture ratio largely below 20% kg/kg. Wall 8 lags behind somewhat. Wall 1, 2, 5, and 6 perform worse. To the north, they still show moisture ratios quite above 20% kg/kg, while to the south they drop just below. The situation in wall 7 is frankly dramatic. There, the high moisture content of the wet sprayed cellulose humidified the studs. Remarkably, due to air looping around and in the insulation moving air from the warm to the cold side at the top, all walls studs dry fastest there. On its way to the bottom, the air cools down causing water vapour picked up at the top to condense down on the sheathing.

Requirement 2

Draping the building paper so the overlaps allow functioning as second drainage plane, avoids rain from wetting the sheathing and timber frame. In addition, overhanging edges mask the delicate façade to roof transition while a backsplash zone in waterproof material above grade is not a redundant luxury with a wood siding or stucco outside finish.

Requirement 3

Requirement 3 determines how to solve the details above grade. In a humid climate, foundation walls and ground floor decks are best executed in a stony material on which the timber-framed walls are mounted. Between grade and lowest bottom plate one must respect a difference in level of at least 20 cm. Also, a continuous damp proof layer should separate the lowest bottom plate from the foundation walls or floor deck. The same damp proof layer is needed everywhere studs contact stony materials that can turn wet.

1.2 Performance evaluation

Requirement 4

Without a continuous air retarder, air-tightness of timber-framed outer walls remains defective. Even when correctly mounted, an air permeance below 10^{-5} kg/(m$^2 \cdot$ s $\cdot$ Pa) at 1 Pa air pressure difference is hardly realizable. As Figure 1.10 underlines, even at moderate air outflow, vapour resistance of the inside finish and building paper have a marginal impact on the amount of condensate deposited at and in the sheathing.

Not only do amounts of condensate vary with height, the worst situation occurs when the leaks at both sides of the insulation are far apart. Clearly, deducing vapour resistance requirements from a Glaser calculation does not work.

Figure 1.11 illustrates the effect of local leaks in the inside finish and the sheathing, coupled to air looping in and around the insulation.

Simulation with more complete models gave following guidance:

1. Construct the envelope as airtightly as possible. Mounting a continuous air barrier foil between thermal insulation and inside finish with a service cavity left is one possibility. An alternative is to air-tighten the inside finish providing perforation afterwards is excluded
2. Thermal insulation must completely fill the space between sheathing and air barrier
3. If 1 and 2 are fulfilled, one must still respect in moderate climates the relations in Table 1.9 between vapour resistance of the air/vapour retarder and vapour resistance of the building paper. For other climates, different relations hold. For example in hot, humid ones that need sensible and latent cooling, the outside finish should have enough vapour retarding quality to exclude high relative humidity and interstitial condensation at the backside of the inside lining.

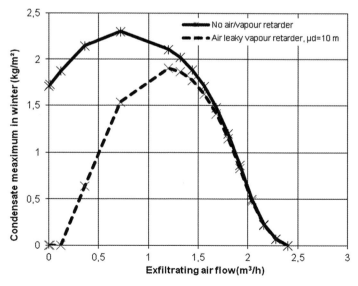

Figure 1.10. Timber-framed outer wall, mineral wool insulated, $U_o = 0.21$ W/(m$^2 \cdot$ K), diffusion thickness of the building paper 0.1 m, indoor climate class 3, moderate Uccle climate: impact of air outflow on maximum condensation deposit against and in the plywood sheathing at the end of the winter.

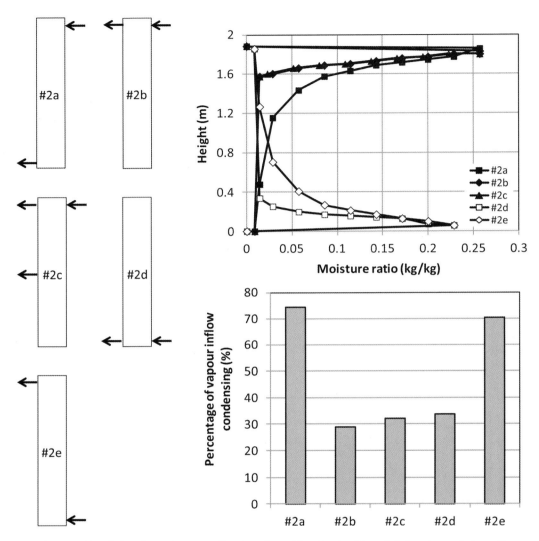

Figure 1.11. Timber-framed outer wall, mineral wool insulated, effect of air looping and air outflow on the distribution of condensate along the wall's height ($\theta_e = -10$ °C, $\theta_i = 20$ °C, $RH_i = 38\%$, outflow: 0.9 m^3/h, situation after 4 days).

Clearly, the requirements in indoor climate class 2 and 3 are far from severe. Or, timber framed outer walls in that type of buildings do not demand excessive vapour tightness at their inside. Air-tightness is what matters.

1.2 Performance evaluation

Table 1.9. Timber framed outer walls, relation between the diffusion thickness of building paper and air/vapour retarder (Uccle moderate climate).

Indoor climate class	Building paper $[\mu d]_{eq}$	Air/vapour retarder $[\mu d]_{eq}$
1	No requirements	
2	$[\mu d]_{air/vapour\ retarder} \leq 1.43$ m and $[\mu d]_{building\ paper} \leq \dfrac{2.6\,[\mu d]_{air/vapour\ retarder}}{2.04 - 1.43\,[\mu d]_{air/vapour\ retarder}}$	
3	$[\mu d]_{air/vapour\ retarder} \leq 2.76$ m and $[\mu d]_{building\ paper} \leq \dfrac{5\,[\mu d]_{air/vapour\ retarder}}{7.62 - 2.76\,[\mu d]_{air/vapour\ retarder}}$	
4, 5	Evaluate per case	

Requirement 5

Surely highly insulated timber framed outer walls finished with a brick veneer may suffer from solar driven vapour flow. An example are passive houses, where the outer walls consist of a timber framed inside leaf, lined inside with an air-tightened OSB sheathing and finished at the cavity side with a very vapour permeable wood fibre board (Figure 1.12). A 3 cm wide unvented cavity separates that inside leaf from a capillary active, 9 cm thick brick veneer, which at the rain side acts as rain buffer storing up to 14 litres per m^2 and more. During warmer weather after a rainy period, part of that moisture diffuses across the inside leaf to the inside where it humidifies the OSB. As the veneer stays at 100% relative humidity year round, relative humidity in the OSB inside lining fluctuates annually as shown in Figure 1.12.

Superimposed is a daily relative humidity oscillation at the OSB's cavity side with peaks over 90% in summer. In fact, temperature at the backside of a wet west over south-west to south looking brick veneer may pass 35 °C during warm summer days. Related vapour saturation pressure then reaches 5260 Pa, high enough to create a daily vapour flow to the inside, which further humidifies the OSB. Solar driven vapour flow activates the OSB's formaldehyde release during the summer months.

Practitioners have no clue of the problems solar driven vapour flow may cause. Avoidance however is simple, as it suffices using building paper that has a slightly higher diffusion resistance than the air/vapour retarding foil or sheathing inside. As Figure 1.13 underlines, such solution fits within the relations of Table 1.9. A less safe alternative consist of ventilating the cavity between brick veneer and timber-framed leaf.

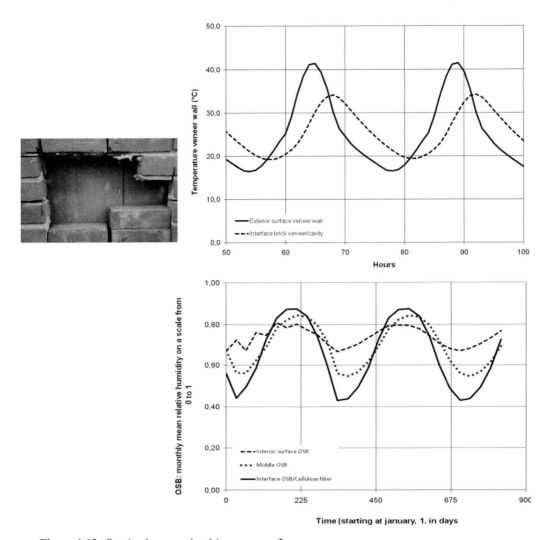

Figure 1.12. Passive house, solar driven vapour flow:
above temperature at the veneers backside, below relative humidity in the inside OSB air retarder.

1.2 Performance evaluation

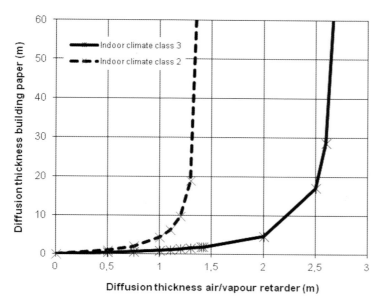

Figure 1.13. Timber-framed outer wall:
relation between the diffusion thickness of the air/vapour retarder and the building paper.

1.2.2.5 Thermal bridges

Limited thermal bridging is a clear advantage of timber-framed construction. Only when very low whole wall thermal transmittances are imposed, does one need engineered studs and alternative solutions for header plates, frame corners, window reveals and lintels, see Figure 1.14. Metal framed construction is a different story. As Table 1.5 showed, correct stud and plate shaping and the use of thermally insulating sheathing then becomes very important.

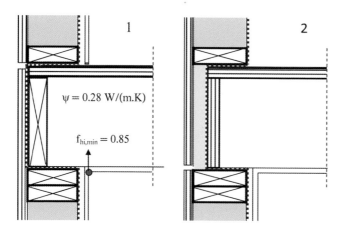

Figure 1.14. Timber-framed outer wall:
adapting header plate design to avoid thermal bridging.

1.2.3 Building physics: acoustics

Timber-framed outer walls insulated with mineral wool or glass fibre have a higher sound transmission loss than windows, though at low frequencies performance fails short. A very high sound insulation, up to $R_w = 52$ dB, demands double walls with at both sides linings with different thickness and the two leafs filled with mineral wool or glass fibre. Party walls between residential units demand that kind of solution (Figure 1.15).

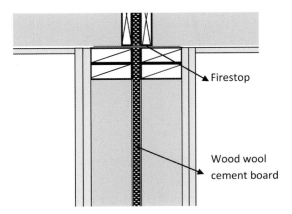

Figure 1.15. Timber-framed party wall.

1.2.4 Durability

Timber is moisture sensitive and deforms anisotropic under hygric load. How to avoid unacceptable wetting is explained above. Platform framing should absorb hygric movements without damage: studs one floor high and sheathing jointed per floor. Each floor deck then acts as a kind of hinge, excluding high hygric movement induced bending moments in the studs.

Figure 1.16. Clapboarding.

Care should be taken with synthetic and timber outside finishes. Synthetics deform thermally, wood hygrically. In both cases the best practice is to use small elements, fastened in a way movement remains possible, as is the case with slated finishes, synthetic siding, aluminium siding, timber siding and timber clapboarding. There, the upper planks cover the nails of the lower ones (Figure 1.16). None of these finishes, however, assures air-tightness. Siding and clapboarding are even not rain-tight, which is why the building paper must be draped in a way it acts as drainage plane.

1.2.5 Fire safety

Timber and timber-framed construction is fairly combustible. Application therefore is only allowed for low-rise construction, up to three storeys, while the inside finish must be fire safe, as gypsum board is. Constructing party walls can anyhow be done in a way, overall fire resistance touches 90′ or more. It suffices to assemble them as sketched in Figure 1.15: two leafs, separated by fire proof wood wool cement boards, the bays between studs filled with mineral wool and both leafs lined with a double layer of gypsum board.

1.2.6 Maintenance

If correctly designed and built – airtight, moisture tolerant, no problematic thermal bridging, hygric movement absorbed without cracking – the maintenance intensity of timber framed hardly differs from massive construction. Of course, maintenance outdoors depends on the finish.

1.3 Design and execution

1.3.1 Above grade

Once the foundation walls are finished, thermally cutting the floor support minimizes thermal bridging to the substructure and ground. Cellular glass blocks are well suited for that (Figure 1.3). After casting the concrete ground floor, the surfaces where the timber framed walls come are levelled. That way the bottom plates are in continuous contact with the deck without needing wedges, a blameworthy practice. Under the bottom plates comes a waterproof layer, and then one anchors these plates with tension bolts into the deck. All joints between plates and waterproof layer are also sealed. An alternative is to use a thick enough polymer bitumen or bitumen pasta as waterproofing.

1.3.2 Frame

Walls are tied together by coupling the single top plates with steel connectors or by using a double top plate with the upper one staggered over the wall's depth (Figure 1.17). For lintels with limited span two sides down mounted studs are used. Larger spans are solved using insulated headers composed of studs and plates with plywood or OSB shearing at both sides. Sometimes a timber ring girder is applied (Figure 1.17). Double studs, of which one acts as jack stud supporting the header, line window and door bays at their sides. An alternative is to fix the headers using steel header hangers.

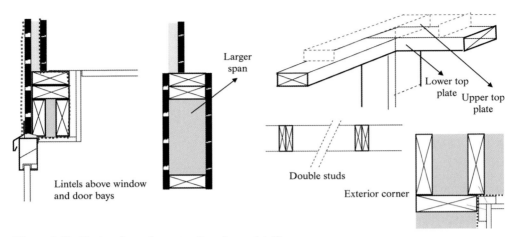

Figure 1.17. Timber-framed construction: frame detailing.

1.3.3 Thermal insulation

Best suited is mineral wool or glass fibre. For so-called sustainability reasons, one also uses sprayed cellulose. Sprayed PUR as alternative guarantees better air-tightness. Anyhow, all bays demand a complete fill between sheathing and airtight layer, which is stapled or mounted after the insulation is put in place.

1.3.4 Air and vapour retarder

As said above, continuity is of prime importance here. The layer also does not tolerate perforation after mounting. Therefore, it is highly recommended to leave a 3 to 5 cm deep service cavity between air and vapour retarding layer and inside lining. Electricity guiding rods and pipes are installed after the air and vapour retarding foil is fixed and the supporting laths for the lining nailed. Then the service cavity is filled with stiff mineral wool boards to guarantee enough mechanical support for the foil not to rip off under wind pressure. Despite all this, it is still common practice to fix all guiding rods in the frame bays before inserting the insulation.

1.3.5 Building paper

We discussed the functions of the building paper above: second drainage plane and additional wind barrier. For more than a decade, spun-bonded foils with good air- and water-tightness but low diffusion thickness, only 0.01 to 0.02 m, have been on the market. A good choice except when a brick veneer or other highly water buffering outer finish is used. In such a case, a control on solar driven vapour flow is a necessity. Anyhow, any building paper foil wraps the sheathing in horizontal stripes, starting at the bottom and going up the envelope with the next strip overlapping the one below over 10 cm or more. Down the building paper, a tray should drain any run-off back to the outside. At door and window bays, the foil is wrapped around the headers and side studs.

1.3.6 Variants

The following variants are quite common: (1) replacing the OSB or plywood sheathing by thermally insulating XPS- or stiff glass fibre boards, (2) composing the construction of prefabricated, modular outside and partition wall timber-framed elements. Variant (1) largely minimizes thermal bridging but in many cases, additional cross bracing tied into the top and bottom plates must guarantee horizontal stiffness.

As said above, timber-framed gained popularity for passive house construction in countries with massive construction tradition. The reason is the ease to insulate with uneconomical thick insulation packages, while keeping thermal bridging minimal. A popular outer wall assembly looks like (from inside to outside):

- Gypsum board inside lining
- Service cavity (not always. If not, all electricity guiding rods, plug sockets, switches and pipes have to be mounted in the partition walls)
- Taped OSB sheathing as air and vapour retarder
- Timber frame using 30 to 35 cm deep engineered studs, the bays filled with mineral wool, or, extremely popular despite its setting, cellulose fibre
- Vapour permeable 0.022 m thick wood wool board sheathing ($\mu = 4.5$)
- Unvented cavity, 3 cm wide
- Outside finish, often a brick veneer

The assembly must guarantee maximum air-tightness and exclude winter interstitial condensation. However, as already shown, the design completely overlooks the negative effects of solar driven vapour flow for walls with brick veneer.

1.4 References and literature

[1.1]   Febelhout, Tekst: *wat is houtskeletbouw, niet gedateerd* (in Dutch).

[1.2]   Latta, J. K. (1973). *Walls, windows and roofs for the Canadian climate.* Special Technical Publication No. 1 of NRC, Division of Building Research, 94 pp.

[1.3]   Architects Journal (1973). Technical Study, Timber.

[1.4]   Trethowen, H. (1976). *Condensation in cavities of building structures.* New Zealand Journal of Science, Vol. 19, 311–318.

[1.5]   STS 23 (1979). Houtbouw (in Dutch).

[1.6]   Laboratorium Bouwfysica (1980). *Invloed van lekken in het dampscherm op het condensatiegedrag van constructiedelen.* Rapport 80/1 (in Dutch).

[1.7]   Mayer, K., Künzel, H. (1980). *Test results concerning the ventilation of the air spaces behind siding using small scale elements.* Report Institut für Bauphysik.

[1.8]   NBI (1981). *Building Research Data Sheets, Timber Frame Walls.* Building Details A523.255.

[1.9]   Samuelson, I. (1981). *Moisture conditions in internally insulated underground walls.* Report CIB-W40, Copenhagen meeting, Swedish National Testing Institute, Borås.

[1.10] NRC (1983). *Division of Building Research, Exterior walls, Understanding the Problems.* Proceedings of the Building Science Forum 1982, Proceedings No. 6.

[1.11] STS 23 (1983). Houtbouw, addendum (in Dutch).

[1.12] Laboratorium Bouwfysica (1983). *Bouwfysische analyse van een woningbouwsysteem*. Rapport 83/38 (in Dutch).

[1.13] Johnson, K. A. (1985). *Interstitial condensation, Building Technical File*. Pilkington Research and Development, No. 8, January.

[1.14] Harderup, L. E., Claesson, J., Hagentoft, C. E. (1985). *Prevention of moisture damage by ventilation of the foundation*. Report Department of Building Technology, Lund Institute of Technology.

[1.15] Laboratorium Bouwfysicsa (1985). *Systeembouw-woning 'Futurhome', Isolatieproblemen*. Rapport 86/5 (in Dutch).

[1.16] Christensen, G. (1985). *Summer condensation in post-insulated exterior walls*. Report CIB-W40, Holzkirchen meeting, Danish Building Research Institute.

[1.17] Vansant, B. (1986). *Hygrothermisch gedrag van houtskeletbouwwanden met hoge thermische prestaties*. Eindwerk KU Leuven (in Dutch).

[1.18] Scanada Consultants Ltd (1986). *Computer model of the drying of the exterior portion of wood framed walls*. Report.

[1.19] Andersen, N. E. (1987). *Summer condensation in an unheated building*. Report CIB-W40, Borås meeting, Danish Building Research Institute, Borås.

[1.20] Bankvall, C. (1987). *Thermal performance of building envelopes with high thermal resistance*. Report, Swedish National Testing Institute.

[1.21] Hens, H. (1987). *Buitenwandoplossingen voor de residentiële bouw: houtskeletbouw*. Intern rapport Laboratorium Bouwfysica (in Dutch).

[1.22] Oefeningen houtskeletbouw (1987–1988, 1988–1989, 1990–1991) (in Dutch).

[1.23] Gockel, H. (1987). *Wood frame construction*. DBZ, 8.

[1.24] McCusaig, L., Stapledon, R. (1987). *A study of the drying potential of various wood-frame systems used in atlantic Canada*. Internal report.

[1.25] Johnson, K. A. (1988). *Improving the thermal performance of timber framed walls*. Building Technical File, Pilkington Research and Development, No. 21, pp. 31–38.

[1.26] Nieminen, J. (1989). *Building Physical Behaviour of a Wooden Wall Structure*. CIB World Congress, Paris.

[1.27] Burch, M. D., Thomas, W. (1991). *An Analysis of Moisture Accumulation in a Wood Frame Wall Subjected to Winter Climate*. NISTIR 4674.

[1.28] CMCH (1991). Air Barrier Technology Update.

[1.29] Proceedings of the ASHRAE/DOE/BTECC Conference on the Thermal Performance of the Exterior Envelopes of Buildings V (1992). ASHRAE, Clearwater Beach.

[1.30] Burch, M. D., TenWolde, A. (1992). *Controlling Moisture in the Walls of Manufactured Housing*. NISTIR 4981.

[1.31] Sandin, K. (1993). *Moisture conditions in cavity walls with wooden framework*. Report Department of Building Materials, Lund University.

[1.32] Lstiburek, J., Carmody, J. (1993). *Moisture Control Handbook: Principles and Practices for Residential and Small Commercial Buildings*. Van Nostrand Reinhold, New York.

[1.33] Hens, H. (1994). *Houtbouw, bouwfysisch en bouwkundig*. Informatiedag 'Bouwen met Hout', Febelhout (in Dutch).

1.4 References and literature

[1.34] Hens, H. (1994). *Lichte Bouwsystemen, een algemene bouwfysische beoordeling.* TI-KVIV studiedag 'Lichte Bouwsystemen' (in Dutch).

[1.35] Information Holz (1994). *Holzbau Handbuch.* Reihe 1, Entwurf und Konstruktion, Teil 3, Folge 2 (in German).

[1.36] Proceedings of the ASHRAE/DOE/BTECC Conference on the Thermal Performance of the Exterior Envelopes of Buildings VI (1995). ASHRAE, Clearwater Beach.

[1.37] Information Holz (1995). *Holzbau Handbuch.* Reihe 1, Entwurf und Konstruktion, Teil 3, Folge 3 (in German).

[1.38] Zarr, R., Burch, D., Fanney, H. (1995). *Heat and Moisture Transfer in Wood-Based Wall Construction: Measured versus Predicted.* NIST Building Science Series 173.

[1.39] Uvslokk, S. (1995). *U-values of wall constructions, influence of workmanship.* Annex 24 report, Norwegian Building research Institute.

[1.40] Künzel, H. (1995). *Regenschutz, Feuchteschutz und Wärmeschutz von Fachwerkwänden, Hinweise für Sanierung und Neubau.* WKSB, Heft 35, pp. 1–9 (in German).

[1.41] MacGowan, A., Desjarlais, A. (1997). *A Investigation of Common Thermal Bridges in Walls.* ASHRAE Transactions, Vol. 103, Part 1.

[1.42] Tuluca, A., Devashish, L., Jawad, H. Z. (1997). *Calculation Methods and Insulation Techniques for Steel Stud Walls in Low-Rise Multifamily Housing.* ASHRAE Transactions, Vol. 103, Part 1.

[1.43] Building Science Corporation (1997). *Builder's Guide: Mixed Climates.*

[1.44] Proceedings of the ASHRAE/DOE/BTECC Conference on the Thermal Performance of the Exterior Envelopes of Buildings VII (1998). ASHRAE, Clearwater Beach.

[1.45] Building Science Corporation (1998). *Builder's Guide: Hot-Dry & Mixed-Dry Climates.*

[1.46] Energy Efficient Building Association (2000). *Builder's Guide: Hot-Humid Climates.*

[1.47] Proceedings of the ASHRAE/DOE/BTECC Conference on the Thermal Performance of the Exterior Envelopes of Buildings VIII (2001). ASHRAE, Clearwater Beach (CD-Rom).

[1.48] Mukhopadhyaya, P., Kumaran, K., Rousseau, M., Tariku, F., van Reenen, D., Dalgliesh, A. (2003). *Application of hygrothermal analyses to optimize exterior wall design.* Research in Building Physics (Eds.: Carmeliet, Hens, Vermeir). A. A. Balkema Publishers, pp. 417–426.

[1.49] Proceedings of the ASHRAE/DOE/BTECC Conference on the Performance of the Exterior Envelopes of Whole Buildings IX (2004). ASHRAE, Clearwater Beach (CD-Rom).

[1.50] Horvat, M., Fazio, P. (2006). *BEPAT-Building envelope performance assessment tool: Validation.* Research in Building Physics and Building Engineering (Eds.: Fazio, Ge, Rao, Desmarais), Taylor & Francis, pp. 277–286.

[1.51] Fazio, P., Rao, J., Alturkistani, A., Ge, H. (2006). *Large scale experimental investigation of the relative drying capacity of building envelope panels of various configurations.* Research in Building Physics and Building Engineering (Eds.: Fazio, Ge, Rao, Desmarais), Taylor & Francis, pp. 361–368.

[1.52] Finch, G., Straube, J., Hubbs, B. (2006). *Building envelope performance monitoring and modelling of West Coast rainscreen enclosures.* Research in Building Physics and Building Engineering (Eds.: Fazio, Ge, Rao, Desmarais), Taylor & Francis, pp. 335–344.

[1.53] Künzel, H., Karagiosis, A., Kehrer, M. (2008). *Assessing the benefits of cavity ventilation by hygrothermal simulation.* Proceedings Building Physics Symposium, Leuven, pp. 17–23.

[1.54] Bassett, M., McNeil, S. (2009). *Drying of insulated and water-managed walls.* Energy Efficiency and New Approaches (Eds.: Bayazit, Manioglu, Oral, Yilmaz), Beysanmatbaa, Turkey, pp. 201–208.

[1.55] Bassett, M., McNeil, S. (2009). *Drying from frame in water-managed walls*. Energy Efficiency and New Approaches (Eds.: Bayazit, Manioglu, Oral, Yilmaz), Beysanmatbaa, Turkey, pp. 208–216.

[1.56] Maref, W., Tariku, F., Di Lenardo, B., Gatland, S. (2009). *Hygrothermal performance of exterior wall using an innovative vapour barrier in Canadian climate*. Energy Efficiency and New Approaches (Eds.: Bayazit, Manioglu, Oral, Yilmaz), Beysanmatbaa, Turkey, pp. 271–278.

[1.57] Proceedings of the ASHRAE/DOE/BTECC Conference on the Performance of the Exterior Envelopes of Whole Buildings X (2007). ASHRAE, Clearwater Beach (CD-Rom).

[1.58] Proceedings of the ASHRAE/DOE/BTECC Conference on the Performance of the Exterior Envelopes of Whole Buildings XI (2010). ASHRAE, Clearwater Beach (CD-Rom).

[1.59] ASHRAE (2011). Handbook of HVAC-Applications, Chapter 44, Atlanta.

2 Sheet-metal outer wall systems

2.1 In general

Sheet-metal outer wall systems are classified as plate, sandwich, or cellular. 'Plate' applies to walls composed of a corrugated inner and outer plate with thermal insulation in between (not so in former times). 'Sandwich' includes all prefabricated modular elements composed of an inner and outer plate with thermal insulation in between. 'Cellular' designates façade walls consisting of vertically or horizontally mounted boxes (Figure 2.1).

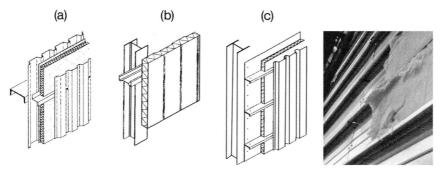

Figure 2.1. (a) plate, (b) sandwich, and (c) cellular metal outer walls.

2.2 Performance evaluation

2.2.1 Structural integrity

Sheet-metal wall systems are non-bearing. That way, any structural evaluation limits itself to a strength and stiffness control under its own weight, wind load and thermal load. Fastening systems of course should be able to transfer the element's own weight and the wind forces to the load bearing building structure, while allowing some movement.

2.2.2 Building physics: heat, air, moisture

Sheet-metal façades face the following problems:

- Lack of air-tightness
- Sensitivity to thermal bridging due to the high thermal conductivity of metals
- Weak transient response. The metal elements are to lightweight for that
- High interstitial condensation risk due to lack of air-tightness

2.2.2.1 Air tightness

The elements themselves are airtight. The problem resides in the joints. These may be unexpectedly air permeable, see Table 2.1 for cellular systems.

Table 2.1. Air permeance of cellular systems ($A = a\, \Delta P_a^{b-1}$).

Cellular system, boxes filled with 80 mm mineral wool	Air permeance $kg/(m^2 \cdot s \cdot Pa)$	
	a	$b-1$
1. No special measures	$7.9 \cdot 10^{-5}$	-0.003
2. Screw eyes caulked	$6.7 \cdot 10^{-5}$	-0.101
3. As 2, joints between boxed taped	$1.6 \cdot 10^{-5}$	-0.084

In many cases the exfiltrating air spreads all over the elements with annoying consequences: clear wall thermal transmittance no longer representative for the true insulation quality, worse transient response than the low weight presumes, disappointing interstitial condensation response, etc.

2.2.2.2 Thermal transmittance

While sandwich elements are mostly thermal bridge free, plate and cellular systems are not. Especially with the latter, thermal bridging deprives the clear wall thermal transmittance of its status as a measure for the insulation performance; see Figure 2.2 and Table 2.2. Due to the large but quite random impact of the contact resistances between the sheet-metal plates, predicting the whole wall thermal transmittance also becomes truly difficult.

Air looping around the thermal insulation might further increase the whole wall thermal transmittance, now called effective thermal transmittance. Filling each box with dense enough joint-free packed mineral wool or glass fibre minimizes looping.

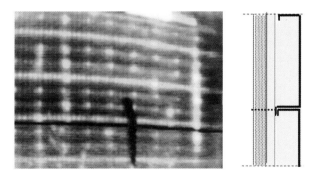

Figure 2.2. Cellular outer wall system.
Left: infrared picture showing thermal bridging, right: optimal thermal cut.

2.2 Performance evaluation

Table 2.2. Cellular outer wall, whole wall thermal transmittance (U_o calculated, U measured).

Cellular wall system, boxes filled with 80 mm mineral wool		U_o W/(m²·K)	U W/(m²·K)	$\Delta U/U_o$ %
1. No thermal cut between box lips and outside plate, which fixed each second lip		0.36	0.74	105
2. No thermal cut between box lips and outside plate, which fixes each lip		0.36	0.89	147
3. No thermal cut between box lips and outside plate, which fixed each second lip, effect of the air layer between lips and outside plate (D in mm)	$D =$ 0.0 0.5 1.0 1.5	0.38	1.10 0.85 0.78 0.74	183 136 116 105
4. Optimal thermal cut between box lips and outside plate (Figure 2.2)		0.36	0.54	50

2.2.2.3 Transient response

Due to their low weight, temperature damping and admittance display disappointing values. Only the dynamic thermal resistance may reach acceptable levels, on condition insulation thickness is large enough, thermal bridging is minimised, air-tightness is guaranteed, and air looping is excluded. In office buildings with sheet-metal outer walls, enough attention should go to moderate glass usage, effective solar shading, floors with assessable heat storage capacity and night-time ventilation, the last only in climates with cool nights.

2.2.2.4 Moisture tolerance

Wind-driven rain

The outside metal surface provides a drainage plane. It should ensure rain-tightness, which is why many plate and cellular façades have vertical plates at the outside, which overall each other vertically away from the prevailing wind direction. For finishing plates with trapezoid profiling, these overlaps are one profile wide. With sandwich façades, the joints are the critical spots. A two-steps solution must be normal practice (Figure 2.3).

Surface condensation

Surface condensation should not be a problem given the insulation thicknesses actually provided. Unfortunately thermal bridging and air looping send that up the wrong way (Figure 2.4). The temperature factor of a cellular wall insulated with 8 cm mineral wool but without thermal cut between box lips and outer sheet-metal plate drops to 0.61 instead of the 0.96 expected in theory. Quite a bit lower than 0.7, the value needed to see surface condensation risk dropping below 5%. Again, an optimal thermal cut gives relief.

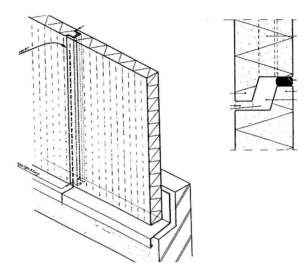

Figure 2.3. Horizontal two-step joint between sandwich elements.

Figure 2.4. Well insulated cellular outer wall without thermal cuts, surface condensation.

Interstitial condensation

As for timber-framed walls, the main cause of interstitial condensation is lack of air tightness. When airtight, diffusion only occurs when the inside plate is non-metallic. Metal has no moisture buffering capacity, which is why interstitial moisture deposit is truly condensation, i.e. droplets form. Further, thermal resistance of the outer sheet-metal plates is low enough so that warming-up by the exfiltrating air does not occur. Therefore, condensate deposited increases proportional to the airflow rate (g_a):

$$g_c = 6.1 \cdot 10^{-6} \, g_a \left(p_i - p_{sat,e} \right) \qquad (2.1)$$

In this equation, p_i is vapour pressure indoors and $p_{sat,e}$ saturation pressure at the sol-air temperature outdoors (all SI). For air exfiltration-coupled interstitial condensation, the indoor climate class so plays the prominent role.

With cellular and plate outer wall systems, interstitial condensation may be a problem. If condensate deposits, it is the backside of the outside sheet-metal that turns wet. Once enough droplets per m² formed, they join and start running off to drip-dry below the finish. This could

2.2 Performance evaluation

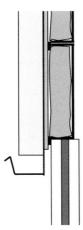

Figure 2.5. Gutter under an outside sheet-metal covering.

be annoying if the finish's underside is above walking height. Frost for example forms the drips into icicles, which break and fall down when thawing starts, a danger for pedestrians. A gutter along the sheet-metal cover clearly is no redundant luxury (Figure 2.5). Besides, the condensate accumulates between sheet-metal and supports. Without rust resisting finish, this will induce corrosion, more so when the condensate is acid (see durability).

Sufficient air-tightness ($g_a < 10^{-5}$ kg/(m² · s) at $\Delta P_a = 1$ Pa) prevents these problems. In climate class 2 and 3, plate and sandwich elements with non-metallic inside finishes also need a class E2 vapour retarder at the inside of the insulation (5 m < $(\mu d)_{eq} \leq 25$ m).

With metal sandwich elements, the joints again are the weak points. If for example the inside seal of a two-steps joint is not airtight, then care should be taken with panels that have the insulation glued against the outside sheet-metal plate. Often continuous air paths as thick as the glue strips remain. Together with the joints they form an area-wide air channel configuration, where in air exfiltration can lead to interstitial condensation (see metal roofs).

2.2.2.5 Thermal bridges

As mentioned, thermal bridging surges where the inside sheet-metal contacts the outside one. In addition to the direct contact as with cellular outer wall systems, the profiles supporting the outer sheet-metal finish, also act as such. For both, contact resistances make the temperature ratio uncertain.

2.2.3 Building physics: acoustics

Sheet-metal outer walls may perform worse than low-e, argon-filled double-glazed windows. If so, investing in sound insulating glass and window makes no sense. First, the performances of the sheet-metal systems must be upgraded: airtight, as little coupling between inner and outer sheet-metal plates as possible, both with different weight and bending stiffness, mineral wool insulation in between, etc. Because sheet-metal façades are typically a solution for industrial buildings, also the inverse problem, protecting the environment against noise produced indoors, demands attention.

2.2.4 Durability

With metals, hygric movement does not occur. However, due to their high modulus of elasticity and a rather significant thermal expansion coefficient, blocked thermal movements can induce high stress. Especially enamelled sheet-metal outside finishes experience large temperature swings, up to 80 °C on an annual basis. Connections must therefore allow limited movement, for example by fastening each plate with a fixed and three pendulum bearings.

The largest problem, however, is corrosion. Electrochemical corrosion for example develops when water runs from metals with higher to metals with lower electrochemical potential, see Table 2.3. The same happens when water fills gaps between two such metals. The main factor driving rusting speed is the local microclimate. Maritime and maritime/industrial environments are most aggravating. An acid, humid inside atmosphere, with 80% relative humidity is a threshold for corrosion to start. Alternating drying and humidification creates the worst circumstance. Aluminium-zinc and coil coating of steel, anodizing and plating of aluminium, polymer protection of zinc, etc., provide efficient protection.

Table 2.3. Metals, electrochemical potential.

Metal	Mg	Al	Zn	Fe	Ni	Sn	Pb	Cu	Ag	Au
Potential (V)	–1.87	–1.45	–0.76	–0.43	–0.25	–0.15	0.13	0.35	0.80	1.50

Low values ⎯⎯⎯⎯⎯⎯⎯⎯⎯⎯⎯⎯⎯⎯⎯⎯⎯⎯⎯⎯→ high values

2.2.5 Fire safety

An overall fire resistance of 30′ must keep fire from spreading to other buildings. With sheet-aluminium plates inside 30′ is hardly attainable. It is with sheet-steel plates.

2.2.6 Maintenance

If correctly designed and mounted (airtight, corrosion protected, etc.), maintenance of sheet-metal outer walls requires only regular cleaning (1 to 2 times a year).

2.3 Design and execution

This subject is too broad to be treated within the frame of this book. We refer to books such as Kettlitz, J. (Ed.) (1994). Handboek voor duurzame metalen gevels en daken (Handbook for durable metal façades and roofs), edited by Hagen en Stam BV, Den Haag (in Dutch), and Koschade, R. (2000). Die Sandwichbauweise (The sandwich building system), edited by Ernst & Sohn, Berlin (in German). They discuss the design and execution of sheet-metal façades and roofs in detail.

2.4 References and literature

[2.1]   NRC, Institute for Research in Construction (1977). *Construction Details for Air tightness.* Record of the DBB Seminar, Proceedings No. 3.

[2.2]   NRC, Institute for Research in Construction (1989). *An Air Barrier for the Building Envelope.* Proceedings of the Building Science Insight '86.

[2.3]   CMHC (1990). Air Barrier Technology Update.

[2.4]   Lecompte, J. (1992). *Binnendoosconstructies, berekend en gemeten (1).* Bouwfysica 3, pp. 24–27 (in Dutch).

[2.5]   de Wit, E. (1992). *Binnendoosconstructies, berekend en gemeten (2).* Bouwfysica 3, pp. 28–31 (in Dutch).

[2.6]   Kettlitz, T. (1994). *Handboek voor duurzame metalen gevels en daken.* Ten Hagen en Stam, Den Haag (in Dutch).

[2.7]   Renckens, G. (1996). *Façades in glas en aluminium.* VMRG, Veenman Drukkers, Wageningen (in Dutch).

[2.8]   Koschade, R. (2000). *Die Sandwichbauweise* (The sandwich building system). Ernst & Sohn, Berlin (in German).

3 New developments

3.1 Transparent insulation

3.1.1 In general

We explained the principle of transparent insulation (TIM) in Performance Based Building Design 1, Chapter 2 on materials. TIM transmits short wave radiation but limits conduction, convection, and long wave radiation. The result is an insulation material that uses solar radiation as efficiently as possible. The boards consist of one glass sheet with the lined or unlined transparent insulation behind or two glass sheets with the insulation in between. There are four forms of application (Figure 3.1):

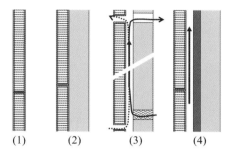

Figure 3.1. TIM-applications: (1) replaces glass, (2) outside insulation, (3) veneer of a ventilated cavity wall, (4) veneer of a ventilated cavity wall with insulated inside leaf.

(1) Replaces glazing (two glass sheets are needed)
(2) Outside insulation for massive walls with the surface behind the TIM-layer painted black (one or two glass sheets)
(3) Veneer of a ventilated cavity wall with the cavity side of the massive inside leaf painted black and the possibility to direct the heated air to the inside or the outside (two glass sheets are needed)
(4) Veneer of a ventilated cavity wall with insulated inside leaf and the possibility to direct the heated air to the inside or the outside (two glass sheets are needed)

3.1.2 Performance evaluation

3.1.2.1 Building physics: heat, air, moisture

Air-tightness

As stated, TIM-elements need an exterior glass protection. Hence, as glass acts as a perfect air barrier, air in- and ex-filtration across the elements is of no concern. However air looping, is. When the TIM does not link up with the exterior glass and an air layer is left at the massive wall side, then the TIM straws and/or the joints between elements allow air looping.

The following example underlines the impact. A 10 cm thick unlined TIM-insulation is sandwiched between an 8 mm thick glass sheet and a 9 cm thick concrete block wall. The air layer width at both sides of the TIM is 20 mm. The joints between the TIM-elements are perfectly closed. Figure 3.2 on top gives the calculated 2D temperature profile in the TIM-wall for an outside temperature of 0 °C, an inside temperature of 20 °C, and no solar gains. The average thermal transmittance touches 1.85 W/(m² · K) and not 0.51 W/(m² · K) as expected without air looping, an increase of 263%! The clear wall thermal transmittance of the concrete block wall without TIM is 2.4 W/(m² · K), only 30% higher. Instead, for a 10 cm thick TIM-layer with lined surfaces and taped joints, temperature profile can be seen as in Figure 3.2 at the bottom. Temperature difference over the TIM now is 14.6°, and the clear wall thermal transmittance equals the expected 0.51 W/(m² · K).

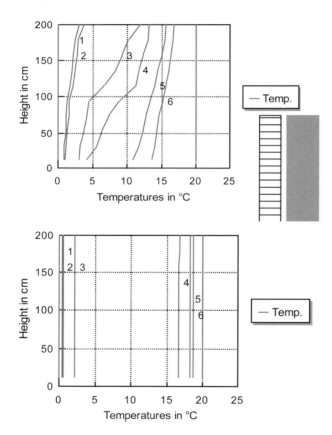

Figure 3.2. Application (2) massive concrete block wall with TIM-boards at the outside, inside temperature 21 °C, outside temperature 0 °C. On top temperatures with, at the bottom without air looping. 1: glass outside, 2: glass inside, 3: TIM outside, 4: TIM inside, 5: concrete outside, 6: concrete inside.

Thermal transmittance

In application (1), the TIM-boards act as light diffusing, somewhat better insulating surfaces. Thermal response is governed by the thermal and solar transmittance. For thicknesses between 11.5 and 40 mm, the thermal transmittance of vertical TIM-panels varies between 2.53 and 1.23 W/(m²·K) at a mean temperature of 10 °C. This is comparable with the difference between normal double-glazing and low-e, argon filled double-glazing. Far from miraculous!

In application (2), (3), and (4), solar radiation transmitted by the TIM heats the massive wall, which even in winter results in monthly mean heat gains for walls looking west over south to east. As a result, not the thermal transmittance but the energy characteristic (E), a number in W/(m²·K) indicating how much heat is gained or lost (E negative) on a monthly or heating season basis per m² and K temperature difference, now figures as efficiency indicator:

$$E = \frac{1}{\left(\overline{\theta}_{i,H} - \overline{\theta}_{e,H}\right) R_{x,i}} \left(\overline{\theta}_{x,H} - \overline{\theta}_{i,H}\right) \tag{3.1}$$

with $\overline{\theta}_{e,H}$ the mean temperature outside, $\overline{\theta}_{i,H}$ the mean temperature inside, and $\overline{\theta}_{x,H}$ the mean temperature of the black painted surface or cavity side of the massive wall. $R_{x,i}$ stands for the thermal resistance between the black painted surface and indoors. In case of application (2), the following heat balance gives the temperature $\overline{\theta}_{x,H}$:

$$\frac{\overline{\theta}_{e,H} - \overline{\theta}_{x,H}}{R_{TIM} + R_e} + \tau_{TIM}\, a_K\, \overline{E}_S + \frac{\overline{\theta}_{i,H} - \overline{\theta}_{x,H}}{R_{x,i}} = 0 \tag{3.2}$$

As an example, Table 3.1 gives the monthly mean energy characteristics for an application (2) wall in the moderate climate of Uccle (Brussels).

Table 3.1. South facing application (2) wall, energy characteristic (10 cm TIM, $U = 0.78$ W/(m²·K)).

Month	J	F	M	A	M	J	J	A	S	O	N	D
E (W/m²·K)	1.79	2.90	3.91	4.70	5.36	5.94	6.74	8.45	7.80	5.39	1.92	1.39
Gain (W/m²)	30	48	61	63	61	56	57	67	74	62	35	23

Even in a moderate climate, heat is gained year round! One must of course interpret the numbers correctly. During cloudy weather, heat is lost but the sunny weather gains are quite beyond the monthly means. On a monthly mean, the gains exceed the losses! Uncontrolled air looping around the TIM-elements may anyhow nullify these gains.

Transient thermal response

A drawback even in moderate climates of the application (1) and (2) is that they induce overheating during the warmer half year. Table 3.2 gives the mean and maximum daily temperature indoors in a living room with floor area 58.8 m², volume 131 m³, envelope area 39.4 m² of which 14.3 m² are windows ($U_{opaque} = 0.22$ W/(m²·K), $U_{window} = 1.8$ W/(m²·K)) during a hot summer day at the end of a heat wave, with outside temperatures touching 31 °C at 2 am (typical for 'hot' days in moderate climates).

Table 3.2. Living room: temperatures during the last hot day of a heat wave.

	Mean temperature °C	Maximum temperature °C
Actual lightweight construction	28.7	34.3
+ outside solar protection	23.1	26.4
+ capacitive construction	21.6	23.4
Actual construction, SW = TIM	33.8	39.8
+ solar protection in the TIM-elements	28.4	33.4

The differences are striking. Possible upgrades are:

a) Inclusion of exterior solar shading in the TIM-elements. Quite effective as Table 3.2 shows but makes the TIM-elements too expensive
b) Application (3). During warm, sunny weather, the outside air ventilating the cavity is sent to the outside
c) Application (4). Equals (3) but the heat conducted across the wall to the inside is minimized thanks to the thermal insulation
d) Applying TIM as in application (2), but embedding a water based heat exchanger in the massive wall, which warms the domestic hot water tank

Testing learned a) and c) is quite effective, whereas b) and d) cause control difficulties and fail in solving overheating.

Moisture tolerance

The discussion only considers application (2) walls.

Rising damp and wind-driven rain

With TIM, rain screening and rising damp demand measures equal to those needed with other enclosure solutions.

Building moisture

The higher temperatures in the massive wall, the presence of an absolute vapour barrier at the outside (the glass) and the glass's very limited solar absorption determine what happens.

The higher temperatures accelerate drying. This will give faster drying of the moist massive wall to the inside, see Figure 3.3, on top. With an interior glass sheet protecting the TIM-boards, some building moisture may also temporarily condense on but not with any deposit in the TIM. If, however, for cost-reasons, that glass is omitted and the massive wall has no vapour barrier finish at its outside, condensate will deposit at the backside of the exterior glass, diminishing the overall solar transmittance that way. Water sucked by the synthetic TIM-straws will raise thermal conductivity, while run off may wet the edge spacers.

Once damp, TIM hardly dries. Solar absorption by the massive wall's black surface in fact has as consequence that even in summer the temperature gradient in the TIM points to the outside, preventing solar driven vapour flow to the massive wall.

3.1 Transparent insulation

After a number of years, the massive wall will anyhow reach hygroscopic equilibrium. For an indoor climate class 3 situation and the Uccle thermal reference year, that equilibrium fluctuates around 35% relative humidity, i.e. a moisture ratio of 1% kg/kg in the concrete blocks. With an opaque outside insulation, the equilibrium is 55% relative humidity, i.e. 1.5% kg/kg.

Interstitial condensation

Compared to an assembly that absorbs most of the incident solar radiation, less solar absorption by the exterior glass sheet tends to give lower temperatures in the most probable condensation interface, the TIM-side of that sheet. Consequently, the probability to see condensate accumulate once the massive wall reaches hygroscopic equilibrium increases in the absence of a vapour barrier at the outside surface of the massive wall (Figure 3.3 at the bottom). A simple analysis shows that nothing less than an absolute barrier is needed, i.e. a glass sheet.

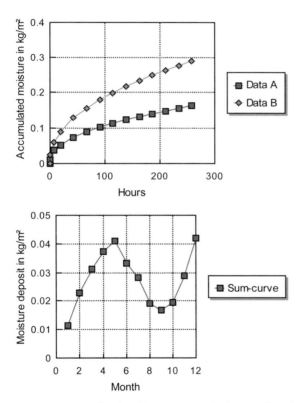

Figure 3.3. Application (2) concrete wall, the TIM-boards without interior glass sheet. Climate class 3 building, inside temperature 21 °C, Uccle outside climate. On top building moisture condensing in the TIM behind the exterior glass sheet (data B) and building moisture drying to the inside (data A). At the bottom interstitial condensation after the concrete reached hygroscopic equilibrium.

Conclusion

The conclusion is straightforward. If an application (2) TIM-wall is not designed correctly, sensitivity to building moisture condensation and interstitial condensation may be far too high. Correct design calls for an absolute vapour barrier in the interface between TIM and massive wall, the best solution being an interior glass sheet and the TIM elements closed with an air- and vapour-tight edge spacer. Even with such hermetically closed elements, the massive wall still must allow some drainage at the backside of the inner glass sheet, though without activating wind washing.

3.1.2.2 Durability

An additional drawback of applications (2) and (3) are higher hygrothermal stresses in the massive wall behind, resulting in increased cracking probability. Experience with TIM since the late 1980s showed that elements, subjected year round to the outside weather conditions in moderate climates, were slowly yellowed by UV-radiation. This caused such a decrease in aesthetic appearance that TIM lost popularity.

3.2 Multiple skin and photovoltaic outer walls

We refer to Chapter 8.

3.3 References and literature

[3.1] Grimme, F. (1993). *Transparent insulation systems: solar improvements for massive walls and windows*. Proceedings IEA ECBCS Future Building Forum workshop on Building Envelope as Energy Systems, Zoetermeer, 13–15/9.

[3.2] Al Bosta, S., Schäfer, K. (1996). *Transparent gedämmte Altbauten, Untersuchung des Einflusses van transparenter Wärmedämmsystemen auf altbauübliche inhomogene Außenwandkonstruktionen*. Forschungsbericht T 2639/3, IRB-Verlag (in German).

[3.3] Lindauer, E., Leonhardt, H. (1994). *Brauchwasservorerwärmung mit transparent gedämmten Bauteilen (Hybridsystem)*. IBP, Mitteilung 246 (in German).

[3.4] Verbeeck, G. (1996). *The importance of radiative heat transfer through transparent insulation material and the effect on the λ-value*. Proceedings of the 8th International Meeting on Transparent Insulation technology, Eurosun '96, Freiburg.

[3.5] Verbeeck, G., Hens, H., Constales, D., Van Keer, R. (1997). *Transparante isolatie: numerieke modellen en evaluatie in situ van de prestaties van het systeem in het VLIET proefgebouw*. REG-potentieel, Jaarverslag IWT (in Dutch).

[3.6] Hens, H., Verbeeck, G. (1997). *Performance assessment of TIM-envelopes*. Proceedings Clima 2000, Brussel.

[3.7] Verbeeck, G., Hens, H., Constales, D., Van Keer, R. (1998). *Transparante isolatie: eindverslag IWT* (in Dutch).

4 Roofs: requirements

4.1 In general

Together with the façades and lowest floor, roofs are part of the building envelope, also called building enclosure. As with outer walls, envelopes separate the human-adapted indoor environment from outdoors. However, more than for outer walls, emphasis with roofs lays on rain control. Indeed, contrary to vertical façades, which only face wind-driven rain, roofs catch all precipitation.

Roofs can be classified according to the cover: rain tight versus watertight. Rain tight indicates the cover figures as rain screen but is not waterproof. Watertight instead means that also water heads are withstood. In such case, we talk about roofing membranes. An alternative is according to roof shape: sloped versus low-sloped. Low-sloped stands for a slope, limited enough not to contribute to the building volume. The discussion in the following chapters assumes shape is not always a relevant criterion. How are canvas roofs inserted? What about sheet-metal roofs, where sloped and low-sloped demand the same technologies?

4.2 Performance evaluation

4.2.1 Structural integrity

Loads bend roofs structures. The often very limited useful load requirements benefit light-weight large span solutions.

That is why space trusses, tie structures, and canvas constructions may be interesting alternatives.

4.2.2 Building physics: heat, air, moisture

4.2.2.1 Air tightness

Roofing is airtight, roof covers are not. For roof assemblies, same requirements as for outer walls prevails: area mean air permeance at an air pressure difference of 1 Pa not exceeding 10^{-5} kg/(m² · s · Pa^b), absence of local air leaks.

4.2.2.2 Thermal transmittance

The whole roof thermal transmittance calculates as:

$$U = U_o + \frac{\sum(\psi_j L_j) + \sum \chi_k}{A_{deel}} \tag{4.1}$$

with U_o the clear roof thermal transmittance, ψ_j the linear thermal transmittance of all linear thermal bridges present, L_j their length and χ_k local thermal transmittance of all local thermal

Table 4.1. Roofs, thermal transmittance, maximum values.

Country	$U_{max} \leq$ W/(m² · K)
Belgium (Flanders)	0.24 (from 2014 on)
Denmark	0.20
Germany	0.20
Finland	0.16
France	0.23–0.30
Luxemburg	0.30
The Netherlands	0.40
Austria	0.35–0.50
UK, Ireland	0.16–0.25

bridges present. The maximum values legally allowed (U_{max}), for some included thermal bridging, differ between countries, see Table 4.1.

The tendency still is further lowering of these thresholds. Yet, perseverance in going that way may end in values beyond the life cycle optimum, which for moderate climates is situated between 0.15 and 0.25 W/(m² · K), depending on the type of insulation material used and the roof construction to be insulated. In passive buildings, values below 0.1 W/(m² · K) are common, i.e. far beyond the optimum.

In low energy buildings, thermal bridge impact should not exceed 0.05 U_o, or:

$$\frac{\sum(\psi_j \, L_j) + \sum \chi_k}{A} \leq 0.05 \, U_o \tag{4.2}$$

The upper threshold for passive buildings is even more severe: 0.01 W/(m · K).

4.2.2.3 Transient response

Simple check

For insulated and energy efficient buildings the dynamic thermal resistance is by definition higher then $1/U_{max}$ m² · K/W with U_{max} the legal thermal transmittance threshold. Temperature damping beyond 15 and thermal admittances exceeding half the inside surface film coefficient ($h_i/2$ W/(m² · K)) are an advantage, not a necessity.

Better check

Dynamic thermal resistance and admittance of roofs must be such that for a given glass type, area, orientation and solar shading, a given infiltration rate, a given ventilation, a given internal heat gain, a given admittance of the partitions and a given dynamic thermal resistance and admittance of all other envelope parts enclosing the space, the number of excess temperature hours (WET-hours) does not pass 100 annually. The number of WET-hours follows from the yearly sum of hourly mean thermal comfort weighting factors WF during the hours of building use:

4.2 Performance evaluation

$$|PMV| \leq 0.5 \quad WF = 0$$
$$|PMV| > 0.5 \quad WF = 0.47 + 0.22\,|PMV| + 1.3\,|PMV|^2 + 0.97\,|PMV|^3 - 0.39\,|PMV|^4$$

with PMV the hourly mean predicted mean vote. Calculations are done for the warm reference year of the location considered.

Anyhow, following simple method guarantees GTO ≤ 100 in a moderate climate without need for a whole year simulation:

- Daily mean temperature indoors (θ_{im}) during a representative hot summer day at the end of a heat wave below or equal to 28 °C
- Daily zone damping $\geq (2.35 - 0.0224\,\theta_{im})/(1 - 0.032\,\theta_{im})$.

The inside temperature θ_{im} follows from a steady state heat balance taking into account all heat flows (transmission, infiltration, ventilation, solar gains, long wave losses to the sky, internal gains) and the sol-air temperature per envelope part, whereas the daily zone damping stands for the harmonic temperature damping of a zone for a complex outdoor temperature with amplitude 1, assuming neither infiltration and ventilation nor solar and internal gains.

4.2.2.4 Moisture tolerance

Roofs are more critical than outer walls. The reasons are a higher rain load, stagnant water creating water heads, risk of dripping condensate, a more pronounced temperature load with more under-cooling and, if low-sloped, more intense solar radiation in summer but less in winter. The following requirements hold:

Water

- *Building moisture*
  Probability the roof turns air-dry without damage within an acceptable period (from one up to a couple of years) beyond 95%.
- *Rain*
  Cover or roofing designed in a way rain penetration is prevented

Water vapour

- *Mould*
  Probability that the mean relative humidity somewhere on the inside surface passes 80% on a four-week basis is less than 5%. In moderate climates this condition is met if the temperature ratio passes 0.7
- *Surface condensation*
  Probability that the relative humidity somewhere on the inside surface of the roof at design temperature equals 100% is less than 5%. In moderate climates this conditions is met for a temperature ratio passing 0.7
- *Interstitial condensation*
  Probability to accumulate moisture deposit is less than 1%. Probability to have yearly returning winter condensate beyond the damage threshold of the material wetted is less than 5%

4.2.2.5 Thermal bridges

For the energy related requirements, see thermal transmittance. In addition, inside temperature ratios anywhere on a thermal bridge should not drop below 0.7. If fulfilled, then, as just stated, probability to get mould or surface condensation is less than 5%.

4.2.3 Building physics: acoustics

As for outer walls, the most important performance is outside noise reduction. Lightweight roofs are also sensitive to contact noise caused by downpour and hail. Assessment in Europe is based on the EN-standards. These express sound transmission loss in dB(A). The value (R_w) follows from shifting the ISO reference curve against the measured or calculated sound transmission loss curve until both coincide on average. Then R_{500}, the sound transmission loss at 500 Hz, read on the shifted reference and corrected for the specific frequency spectrum of the noise (traffic or other sources), quantifies the wall's performance (Figure 4.1).

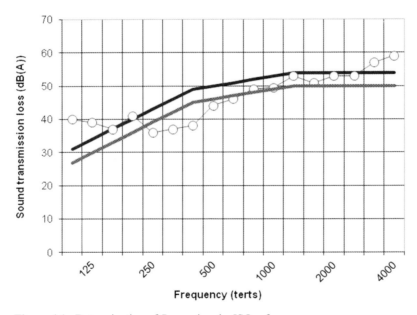

Figure 4.1. Determination of R_{500}, using the ISO reference curve.

4.2.4 Durability

For roofs, hygrothermal stresses create more problems than they do for outer walls. Indeed, the hygrothermal load is higher and the consequences of tear and shear the more devastating.

4.2.5 Fire safety

Medium and high-rise buildings use their low-sloped roofs as escape routes. Therefore, the roofing must have the correct fire reaction (hardly burnable, not contributing to flame spread) and the load bearing structure must guarantee a high enough fire resistance against collapse. In general the demands are:

	Number of storeys		
	≤ 2	3 to 5	> 5
Fire class	C	C	B
Fire resistance			90′

4.2.6 Maintenance and economy

Low-sloped roofs demand regular maintenance, with cleaning fallen leafs as one of the duties. From an economical point of view, life cycle costs are the reference.

4.3 References and literature

[4.1] Becker, R., Paciuk, M. (1996). *Application of the Performance Concept in Buildings.* Proceedings of the 3th CIB-ASTM-ISO-RILEM International Symposium, Vol. 1 and 2.

[4.2] Hendricks, L., Hens, H. (2000). *Building Envelopes in a Holistic Perspective, Methodology.* Final Report IEA-ECBCS Annex 32 'Integral Building Envelope Performance Assessment', ACCO, Leuven.

5 Low-sloped roofs

5.1 Typologies

Low-sloped roofs normally have an impervious membrane on top. Imperviousness also stands for vapour-tightness, meaning that in moderate and cold climates the vapour barrier is on the wrong side. This quandary dominated practice for a long time. Two low-sloped roof types surfaced: the non-ventilated, also called compact or 'warm', and the ventilated, sometimes named 'cold'. With the compact one, inclusion of a vapour barrier aims at preventing water vapour permeating from inside to condense in the thermal insulation. Ventilation was intended to remove water vapour diffusing into the assembly from indoors before it condensed against and in the roof boarding. In the 1970s, a third type surfaced: the protected membrane roof.

All material layers in compact roofs link up with each other: inside finish, load-bearing deck, thermal insulation and membrane (Figure 5.1). The layer under the membrane is called the substrate. At first sight, except for the membrane, the layer sequence looks free. Figure 5.1 is a possibility. Anyhow, in most cases, somewhere in the assembly a layer should be located that ensures a 1.5% fall to the downspouts, while a vapour barrier must be included if necessary. It is at the winter warm side of the insulation, but where is a questions mark, the other being the correct layer sequence. Also in protected membrane roofs, layers link up with each other. The difference with the compact type is where the membrane sits, now below the insulation (Figure 5.2).

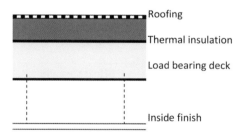 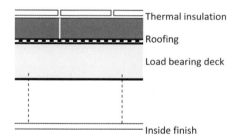

Figure 5.1. Compact low-sloped roof.   **Figure 5.2.** Protected membrane roof.

The substrate in ventilated roofs was separated from the other layers by an outside air ventilated cavity, giving as assembly down-up: inside finish, load bearing deck, thermal insulation, ventilated cavity, boarding (forms the substrate), membrane. Abundant field experience (Figure 5.3) and several experiments during the 1970s catalogued the roof type as extremely risky in terms of moisture response. Not only did designers abuse the safe feeling outside air ventilation gave, by proposing all kinds of incompetent assemblies but the redundancy in measures needed to limit damage risk was such that the investment per m^2 turned out much higher than for compact roofs. Consequently, the ventilated type was classified in good practice guides beginning in the mid-1980s as 'don't apply'.

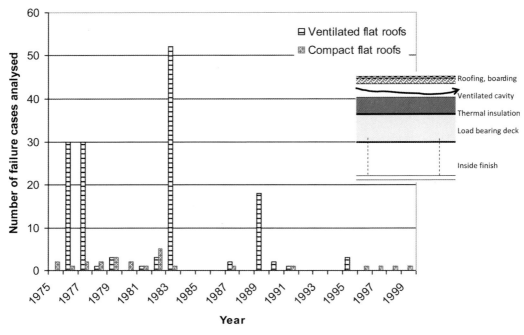

Figure 5.3. Ventilated low-sloped roof, damage cases investigated from 1975 on.

Further classification of compact and protected membrane roofs is based on the type of load bearing deck. One distinguishes between:

Roof type	Load bearing deck		
	Lightweight	Semi-heavy weight	Heavy weight
Compact	X	X	X
Protected membrane		X	X

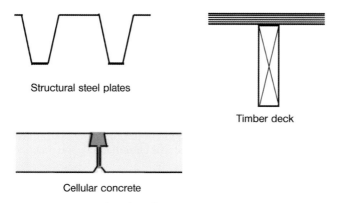

Figure 5.4. Low-sloped roofs: lightweight and semi-heavy decks.

5.2 Roofing membranes

'Lightweight' includes structural sheet-metal plates as well as timber rafters with plywood, particleboard, cement fibreboard or OSB as walk-on finish. In several cases, semi-heavy decks, such as structural cellular concrete floor units, had a thermal resistance high enough to make extra insulation redundant. 'Heavy' includes all massive decks that have a too low thermal resistance to walk along without thermal insulation: reinforced concrete slabs, concrete ribbed floor slabs, structural concrete floor units (see Figure 5.4)

5.2 Roofing membranes

As stated, rain tightness is a requirement any roof has to fulfil. With low-sloped roofs, precipitation may create pools, resulting in water heads, at a 100 Pa per cm water gauge. For that reason, avoiding leakage demands impervious roofing membranes. In volume 1, we presented the membrane types: bituminous, polymer bituminous and polymer.

5.2.1 Build-up, multi-ply roofing

Actual multi-ply roofing solutions combine bituminous with polymer bituminous membranes. Indeed, bituminous top membranes on well insulated low-sloped roof assemblies age so fast only polymer bitumen guarantees acceptable service life. Variables characterizing multi-ply build-up roofing are: (1) number of layers, (2) execution, (3) bonding, (4) protection.

5.2.1.1 Number of layers

The choice is between single coverage and two-ply solutions. Two-ply provides much smaller leakage risk and is less damage sensitive. What to choose anyhow depends on the building's function and expected maintenance intensity. If function requires a very low leakage risk and/or if poor maintenance is expected, apply two-ply. In all other cases, single coverage is a fair choice.

- *Two-ply*
  Consists of a base and top layer. A bitumen or polymer bitumen membrane forms the base layer, while top layers are polymer bitumen (4 mm SBS or APP, Figure 5.5).
- *Single coverage*
  Consists of a polymer bitumen (4 mm SBS or APP).

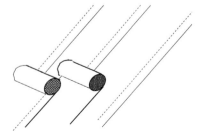

Figure 5.5. Two-ply.

5.2.1.2 Execution

Choices are loose-laid, fully bonded, partially bonded or mechanically fastened.

- *Loose laid*
  Base or single coverage layer are separated from the substrate. To avoid bonding afterwards, one first lays a glass fibre or felt interlayer. The finished roofing gets ballast. Independent movement of substrate and roofing is an advantage, but the fact that wind suction cannot be transmitted to an air permeable roof construction is a drawback.
- *Full bond*
  Base or single coverage layer is bonded to the substrate over its entire area. That wind suction transmits to the roof construction without causing important stresses in the roofing is an advantage, but the fact that each movement of the substrate deforms the roofing is a drawback.
- *Partial bond*
  The base layer, a perforated membrane, is bonded point wise to the substrate by using a perforated membrane as base layer (Figure 5.6). The fact that substrate movement spreads over larger widths in the roofing is an advantage, but the fact that wind suction induces large stress concentration in all bonding dots is a drawback.
- *Mechanically fastening*
  Demands a two-ply execution. Only the base layer is fastened to the substrate using countersunk screws. The top layer is fully bonded to the base layer. This way of executing demands a mechanically strong substrate. For the advantages and drawbacks, see partially bonding.

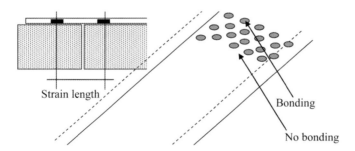

Figure 5.6. Partial bond.

5.2.1.3 Bonding

Choices are: adhering, sealing or sealing with cold glue:

- *Adhering*
  The ply's are adhered mutually and on the substrate with hot bitumen, pro rata of 1 l/m^2 for non- and 1.5 l/m^2 for perforated bitumen.
- *Sealing*
  Layers are adhered mutually and on the substrate using a gas burner.
- *Sealing with cold glue*
  Layers are adhered mutually and on the substrate with bitumen or PU-based cold glue.

5.2 Roofing membranes   55

5.2.1.4 Protection

Used are heavy ballast and slate chippings.

- *Heavy ballast*
  The top layer or single coverage gets gravel ballast or button tiles as protection. Both screen UV and withstand wind suction. When poorly maintained, moss and plants may grow on and in the gravel or at the joints between tiles.
  Loose laid roofing requires heavy ballast.
- *Slate chipping*
  UV screening is taken up by slate chippings enrolled in the top layer during manufacturing.

5.2.1.5 Combination with the type of substrate

The kind of substrate co-defines which roofing solution to use:

Substrate	Choices
Two-ply	
Insulation materials	
Thermally stable, low tensile strength Dense mineral wool or glass fibre boards Perlite boards	The low tensile strength excludes partial bonding. Options are: • Base layer loose-laid or fully bonded • Top layer adhered, sealed or cold glued • Heavy ballast when base layer is loose laid, slate chipping if fully bonded. If for one or the other reason no ballast can be used and full bonding is impossible, the base layer must be mechanically fastened across the insulation into the roof deck
Thermally stable, high tensile strength Cellular glass	All types are applicable except mechanical fastening: • Base layer loose laid or fully bonded • Top layer adhered, sealed or cold glued • Heavy ballast if base layer is loose laid, slate chipping if fully bonded
Synthetic foams EPS, XPS, PUR	Due to the important thermal movements, a full bond is not applicable. Options are: • Base layer loose laid or partially bonded. An alternative is to mechanically fasten the base layer across the insulation into the roof boarding or roof deck • Top layer adhered, sealed or cold glued • Heavy ballast when base layer is loose laid, slate chipping if fully bonded
Timber, cellular concrete, concrete	
Due to their hygroscopicity install the base layer in a way vapour pressures can redistribute over the roof area	
Timber	• Base layer mechanically fastened • Top layer adhered, sealed or cold glued
Cellular concrete	• Base layer loose laid, partially bonded or mechanically fastened • Top layer adhered, sealed or cold glued • Heavy ballast when base layer is loose laid, slate chipping if partially bonded or mechanically fastened
Concrete	• Base layer loose laid or partially bonded • Top layer adhered, sealed or cold glued • Heavy ballast when base layer is loose laid, slate chipping if partially bonded

Substrate	Choices
Single coverage	
Insulation materials	
Thermally stable, low tensile strength	• Loose laid or fully bonded • Heavy ballast when base layer is loose laid, slate chipping if fully bonded
Thermally stable, high tensile strength	• Loose laid or fully bonded • Heavy ballast when base layer is loose laid, slate chipping if fully bonded
Synthetic foams	• Loose laid or partially bonded • Heavy ballast when base layer is loose laid, slate chipping if fully bonded
Timber, cellular concrete, concrete Due to their hygroscopicity install the base layer in a way vapour pressures can redistribute over the roof area	
Cellular concrete	• Partially bonded • Slate chipping
Concrete	• Loose laid or partially bonded • Heavy ballast when base layer is loose laid, slate chipping if partially bonded

Heavy ballast weighs some 120 kg/m^2. When structurally designing the roof, this has to be accounted for. Before partially bonding the base layer, one first points cellular concrete and concrete with bituminous varnish. The base layer or single covering then is rolled out parallel to the falling gradient and loose laid, partially or fully bonded. Aside from that, the overlaps between rolled strips must always be bonded carefully. When adhering the top to the base layer in two-ply applications, the overlaps between rolled strips also demand careful bonding. The top layer should also shift half a rolled-strip width compared to the base layer (Figure 5.5). Once the surface is finished, the roof edges and set on edges get separately bonded rolled-strips that overlap the top layer along the sloped or vertical edge plane.

5.2.2 Build up polymer roofing

Polymer roofing is single coverage. The membranes are too expensive for two-ply application. The largest problem is overlaps. To waterproof, different techniques are available. With thermoplastics, overlaps are welded with hot air or fused together using solvents. With thermo-hardenings, a much-applied technique is gluing. Applying polymer roofing anyhow demands a specialized workforce. Most system manufacturers provide for their instruction.

5.2.3 Problems with roofing

Following problems surfaced: (1) pimples, (2) alligator skin, (3) cracking, (4) blistering.

5.2.3.1 Pimples

Pimples are little bubbles arising in the top membrane caused by a wet organic insert. When sun lighted, vapour saturation pressure there reaches high enough values to lift the warm bitumen. At night, undercooling stiffens the bitumen again, so the bubbles formed keep their shape. Repetitive sun lighting, followed by undercooling allows them to grow. When walked over, they break, creating little craters that expose the insert.

5.2 Roofing membranes

5.2.3.2 Alligator skin

The wording 'alligator skin' could be taken literally. After a number of years, bituminous top layers start showing a pattern of merging hexagonal cracks. The cause is hardening and shrinkage by high temperatures and UV radiation.

5.2.3.3 Cracking

Cracking splits two-ply roofing across the whole thickness. Bitumen membranes with glass fibre insert are very sensitive to that. Cracking happens where two-ply roofing experiences large cyclic deformations of the substrate, for example above joints between synthetic insulation foams, see Figure 5.7. In addition, water pools on the roofing, which expand when freezing, facilitate cracking.

Figure 5.7. Cracking.

5.2.3.4 Blistering

Blistering differs from pimples by the size of the bubbles: large instead of small (Figure 5.8). Also the cause is not a wet insert but locally bad bonding between top and base membrane, enclosure of air and moisture in between and application on a tight substrate. Driving force is the succession of insolation and cooling. High temperatures increase air pressure in the badly bonded spots. This and the high vapour saturation pressure due to the moisture present, inflates the warm, soft bitumen top membrane a little. At night the cold top layer hardens again, so that the under-pressurized bubble cannot regain its original shape, sucking air instead. In each cycle, this process recurs, allowing the blisters to grow. Although blistering does not cause leakage, it impedes drainage. Pond formation and drying accelerates aging, while the bitumen on top of the blisters slowly thins by yield. Blisters break and leakage starts when walked over during cold weather.

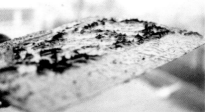

Figure 5.8. Blistering.

5.3 Compact low-sloped roofs

5.3.1 Assemblies

5.3.1.1 Heavy deck

Since insulation is truly necessary but a vapour retarder perhaps not, a reference section contains at least five layers: inside finish, load-bearing deck, thermal insulation, sloping screed and roofing membrane. That results in three variants (Figure 5.9):

Assembly H 1.0	Assembly H 2.0	Assembly H 3.0
Roofing membrane	Roofing membrane	Roofing membrane
Sloping screed	Sloping screed	*Thermal insulation*
Deck	*Thermal insulation*	Sloping screed
Thermal insulation	**Deck**	**Deck**
Inside finish	Inside finish	Inside finish

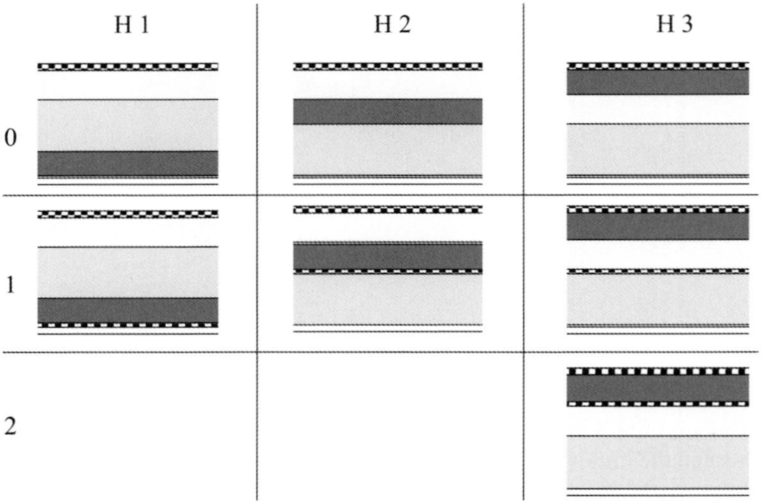

Figure 5.9. Heavy deck: matrix with all variants.

If a vapour barrier is needed, four variants add:

Assembly H 1.1	Assembly H 2.1	Assembly H 3.1	Assembly H 3.2
Roofing membrane	Roofing membrane	Roofing membrane	Roofing membrane
Sloping screed	Sloping screed	*Thermal insulation*	*Thermal insulation*
Deck	*Thermal insulation*	Sloping screed	Vapour barrier
Thermal insulation	Vapour barrier	Vapour barrier	Sloping screed
Vapour barrier	**Deck**	**Deck**	**Deck**
Inside finish	Inside finish	Inside finish	Inside finish

Also see Figure 5.9. Because a vapour barrier at the underside of the deck lacks buildability such an assembly is not included.

5.3 Compact low-sloped roofs

5.3.1.2 Semi-heavy deck

Usually the load-bearing deck gets a little gradient, making a sloping screed redundant. In addition, a thermal insulation is often not needed. Then, only the inside finish (not always), the load-bearing deck and the roofing membrane remain, i.e. two or three layers. This gives variant SH 0.0. If an insulating layer cannot be missed, the variants SH 1.0 and SH 3.0 add. A vapour barrier introduces four extra variants: SH 0.1, SH 1.1, SH 3.1 and SH 3.2

Assembly SH 0.0	Assembly SH 1.0	Assembly SH 3.0
Roofing membrane	Roofing membrane	Roofing membrane
Deck	**Deck**	*Thermal insulation*
(Inside finish)	*Thermal insulation*	**Deck**
	Inside finish	Inside finish

Assembly SH 0.1	Assembly SH 1.1	Assembly SH 3.1	Assembly SH 3.2
Roofing membrane	Roofing membrane	Roofing membrane	Roofing membrane
Deck	**Deck**	*Thermal insulation*	*Thermal insulation*
Vapour barrier	*Thermal insulation*	**Deck**	Vapour retarder
Inside finish	Vapour retarder	Vapour retarder	**Deck**
	Inside finish	Inside finish	Inside finish

5.3.1.3 Light-weight deck

Again, the load-bearing deck gets some gradient. As thermal insulation anyhow is a necessity, four layers remain: inside finish, load-bearing deck, thermal insulation and roofing membrane. This gives two assemblies: L 1.0 and L 3.0. If a vapour barrier is a must, then L 1.1 and L 3.1 add. L 1.1 is possible on condition lined or vapour tight insulation boards with air-tightened joints are used.

Assembly L 1.0	Assembly L 1.1	Assembly L 3.0	Assembly L 3.1
Roofing membrane	Roofing membrane	Roofing membrane	Roofing membrane
Deck	**Deck**	*Thermal insulation*	*Thermal insulation*
Thermal insulation	*Thermal insulation*	**Deck**	Vapour retarder
Inside finish	Vapour barrier	Inside finish	**Deck**
	Inside finish		Inside finish

5.3.1.4 Conclusion

The number of possible and buildable variants scores high, eighteen! A performance evaluation must determine which win. The most important questions are: how much thermal insulation, and where, whether vapour barrier is needed or not, and where to install it?

5.3.2 Performance evaluation

5.3.2.1 Structural integrity

The load-bearing deck of low-sloped roofs provides strength and stiffness. Yet, the ratio between dead load and weight at one side and useful load at the other differs from floors. When a roof is only accessible for maintenance, useful load equals 1000 N/m². In case the inhabitants may use the roof as terrace, useful load becomes 2000 N/m². Wind creates suckion. How much depends on the air-tightness of roof and façades and the location looked for: higher at corners and edges than at the centre, see Table 5.1 and Figure 5.10. With lightweight low-sloped roofs, suction can overhaul the deck's weight and dead load. This requires anchoring the deck in the load bearing structure. Standards and good practice guides give additional data on wind load.

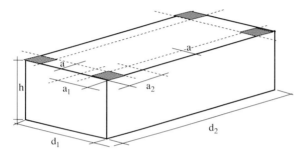

Figure 5.10. Low-sloped roofs: wind suction, edge and corner zones.

Heavy windstorms can tear off lightweight roofs and pile up the insulation boards (Figure 5.11). This however only happens when the deck is air permeable. When airtight, suction ($= \Delta V$) causes under-pressure between membrane and deck:

$$\Delta P_a = -P_a \frac{\Delta V}{V} \tag{5.1}$$

Figure 5.11. Lightweight roof, membrane torn off during a wind storm.

This way equilibrium sets in again, preventing the membrane from being torn off. Of course, besides air-tightness, wind dynamics also intervene. Lightweight decks may resonate under pulsating wind. Vibration amplitude then generates important inertia forces on membrane and thermal insulation, with sometimes devastating consequences.

5.3 Compact low-sloped roofs

Table 5.1. Wind suction on low-sloped roofs.

Location	Height above grade of the roof									
Coast	–	–	–	–	–	–	7.0	9.0	11.5	14.5
Country side	–	5.0	6.0	7.5	9.5	12.0	14.0	18.0	22.0	27.0	32.0
Suburban	5.0	11.0	13.0	16.0	19.0	23.0	27.0	32.0	40.0	46.0	54.0
City centre	18.0	19.5	22.0	26.0	32.0	37.0	42.0	50.0	57.0	66.0	78.0
Wind suction (Pa)										
Façades and roof air permeable										
Corners[3] Low rise[1]	2089	2165	2310	2475	2640	2805	2970	3135	3300	3465	3630
High rise[2]	1772	1837	1960	2100	2240	2380	2350	2660	2800	2940	3080
Edges[3] Low rise[1]	1772	1837	1960	2100	2240	2380	2350	2660	2800	2940	3080
High rise[2]	1456	1509	1610	1725	1840	1956	2070	2185	2300	2415	2530
Centre[3]	1139	1181	1260	1350	1440	1530	1620	1710	1800	1890	1980
Façades air-tight, roof air permeable										
Corners[3] Low rise[1]	1772	1837	1960	2100	2240	2380	2350	2660	2800	2940	3080
High rise[2]	1456	1509	1610	1725	1840	1956	2070	2185	2300	2415	2530
Edges[3] Low rise[1]	1456	1509	1610	1725	1840	1956	2070	2185	2300	2415	2530
High rise[2]	1139	1181	1260	1350	1440	1530	1620	1710	1800	1890	1980
Centre[3]	823	853	910	975	1040	1105	1170	1235	1300	1365	1430
Façades and roof air-tight										
Corners[3] Low rise[1]	1266	1312	1400	1500	1600	1700	1800	1900	2000	2100	2200
High rise[2]	950	984	1050	1125	1200	1275	1350	1425	1500	1575	1650
Edges[3] Low rise[1]	950	984	1050	1125	1200	1275	1350	1425	1500	1575	1650
High rise[2]	633	656	700	750	800	850	900	950	1000	1050	1100
Centre[3]	317	328	350	375	400	425	450	475	500	525	550

[1] Low rise: ratio between height and largest plan view dimension below or equal to 1
[2] High rise: ratio between height and largest plan view dimension beyond 1
[3] Corners, edges, see Figure 5.10

Width of the edge zone (a):

$h \geq \dfrac{d_1}{3}$	$h < \dfrac{d_1}{3}$
$a = \max(0.15\, d_1, 1)$ in m	$a = \max(0.45\, h, 0.04\, d_1, 1)$ in m

Length of the corner zones (a_1 and a_2):

$d_2 \geq 1.5\, d_1$	$d_1 < d_2 < 1.5\, d_1$
$a_1 = a$	$a_1 = 0.5\, d_1 \left(1.5 - \dfrac{d_2}{d_1}\right) + a \left(\dfrac{d_2}{d_1} - 0.5\right)$
$a_2 = 0.5\, d_1$	$a_2 = 0.5\, d_1 \left(\dfrac{d_2}{d_1} - 0.5\right) + a \left(1.5 - \dfrac{d_2}{d_1}\right)$

5.3.2.2 Building physics: heat, air, moisture

Air-tightness

In principle, the membrane should provide air-tightness. But it may not work. Take the roof of Figure 5.12: plywood deck with air-permeable joints between boards, 10 cm dense mineral wool as thermal insulation, a vent with cross section 28 cm² per 20 m² of membrane, no vapour retarder.

The vents should provide vapour pressure relief under the membrane. According to Glaser, a vapour retarder is redundant in indoor climate class 2 buildings (no annual accumulating condensate, acceptable annual maximum). For the given roof however, a diffusion-based calculation is inappropriate. Due to the vents, the membrane lacks air-tightness, as do the plywood deck and the insulation. The consequence is air exfiltration. How much, depends among others on the leakage across the façade walls. If these are quite air permeable, then, as Figure 5.12 shows, one gets much more condensate than from diffusion. Or, the roof requires an air barrier. A correctly adhered vapour retarder fulfils that function to a tee.

Also, the edges may compromise air-tightness, as is the case with a structural sheet-metal deck.

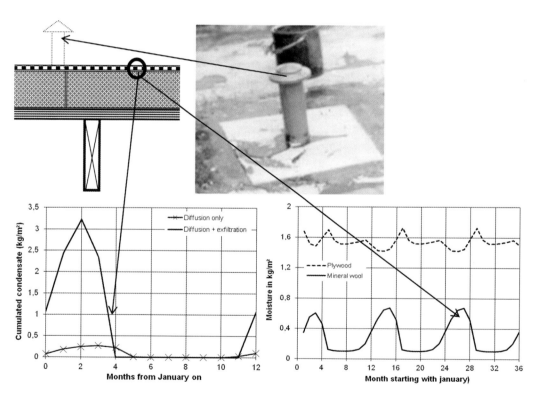

Figure 5.12. Interstitial condensation in a lightweight compact roof, no vapour barrier, ICC 2. The graph on the left gives the condensate deposit by diffusion (Glaser, thin line). On the right the same, now calculated with Match®. Moisture deposit in the mineral wool surpasses the Glaser data. The thick line in the graph on the left adds exfiltration (air permeance coefficient $1.2 \cdot 10^{-4}$ kg/(m² · s · Pa$^{0.56}$)). The difference in deposit is huge.

5.3 Compact low-sloped roofs

Thermal transmittance

Heavy deck

Starting-point is a low-sloped roof with a 14 cm reinforced concrete deck and a sloped screed in lightweight concrete, which on average is 12.5 cm thick, with a maximum of 20 and a minimum of 5 cm. A correct calculation accounts for that change in thickness:

$$U_o = \frac{1}{R_1} \ln\left(1 + \frac{R_1}{R_{a,o}}\right) \tag{5.2}$$

with R_1 thermal resistance of the sloped screed with as thickness the difference between minimum and maximum value and $R_{a,o}$ thermal resistance environment to environment of the roof included the sloped screed at minimum thickness.

As limits for the clear thermal transmittance we take $0.1 \leq U_o \leq 0.4$ W/(m²·K), 0.1 being common in passive buildings. This demands the insulation thicknesses of Table 5.2 (±0.5 cm), whatever the location of the layer in the assembly may be. In In other words, no differences exist between the assemblies H 1.0, H 2.0 and H 3.0. XPS and EPS are not included in the table. Both materials should be discouraged from application directly under the membrane. If they are applied, their limited temperature stability requires heavy ballast, while the EPS boards are best lined up with bitumen glass fibre at both sides.

A value 0.2 W/(m²·K) does not demand extreme thicknesses although the edges and all set on edges must be heightened to allow for the extra centimetres. With 0.1 W/(m²·K), only PUR/PIR gives an acceptable thickness. The other three demand such edge and set on edge heights that the additional cost may exceed the life cycle optimum.

Table 5.2. Heavy deck: insulation thicknesses.

Insulation material	U_o (W/(m²·K))			
	0.4	0.3	0.2	0.1
	Thickness in cm			
Dense mineral wool boards	7.5	11	17	34
Dense glass fibre boards	7.5	10	16	33
Cellular glass	10	14	21	43
PUR/PIR	5	7	11	22

Semi-heavy deck

Reference here is prefabricated cellular concrete deck elements. The thickness needed depends on the span to be bridged. Manufacturers typically deliver the elements 15, 20 and 28 cm thick. Air-dry, clear thermal transmittance is:

Thickness cm	U_o W/(m²·K)
15	0.87
20	0.68
28	0.51

Even with a thickness of 28, a value 0.4 W/(m² · K) is out of reach. Reality is even less favourable. Fresh cellular concrete contains high manufacturing moisture content, giving an initial clear roof thermal transmittance, equal to:

Thickness cm	$U_{o,\text{building moisture}}$ W/(m² · K)
15	> 1.35
20	> 1.09
28	> 0.84

If the objective is a value below 0.4 W/(m² · K) from the first year on, extra insulation is needed, see SH 1.0 and SH 3.0. This way, the choice loses simplicity. Therefore, prefabricated cellular concrete deck elements mainly find application in industrial premises, where the energy performance requirements are less strict.

Light-weight deck

Let us return to the roof of Figure 5.12. With a gypsum board ceiling, a clear thermal transmittance 0.4 to 0.1 W/(m² · K) demands the insulation thicknesses of Table 2.3 (±0.5 cm), hardly different from those for heavy deck roofs and equal for assembly L 1.0 and L 3.0.

Table 5.3. Light-weight deck, insulation thicknesses.

Insulation material	U_o (W/(m² · K))			
	0.4	0.3	0.2	0.1
	Thickness in cm			
Dense mineral wool boards	6	9	15	33
Dense glass fibre boards	6	9	15	31
Cellular glass	8	12	20	41
PUR/PIR	4	6	10	20

A value of 0.2 W/(m² · K) again does not demand extreme thicknesses though the edges and all set on ones must also here be heightened to allow the extra centimetres. With 0.1 W/(m² · K), only PUR/PIR gives an acceptable thickness. The other three demand such extra edge and set on edge heights that the additional cost may exceed the life cycle optimum.

Thermal bridge impact

The more strict the insulation requirement, the more thermal bridging widens the gap between whole and clear thermal transmittance. Edges, set on edges and roof superstructures (domes, chimneys, lift shafts, rising walls) are critical. Edge solutions demand continuity between outer wall and roof insulation. Figure 5.13 illustrates this for a heavy weight compact roof.

5.3 Compact low-sloped roofs

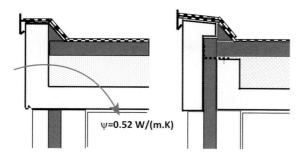

Figure 5.13. Heavy roof, edges.

The high linear thermal transmittance of the roof edge on the left presumes important thermal bridging. The roof edge on the right performs much better. Roof and outer wall insulation connect, while the roof insulation remains in plane. Figure 5.14 shows an edge of a light-weight compact roof. On the left, cavity closure creates problems. On the right, prefabrication allows construction of a thermal bridge-free edge.

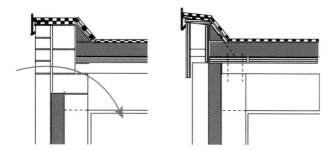

Figure 5.14. Lightweight roof, edges.

Careless detailing of underpinning beams also fosters thermal bridging, as shown in Figure 5.15. Permanent insulating formwork at the inside is no solution. Only a drastic adjustment, which restores continuity between wall and roof insulation, gives relief. Figure 5.16 shows how to create roof domes that are thermal bridge-free.

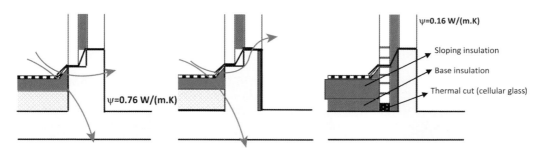

Figure 5.15. Heavy compact roof, underpinning beam, on the left careless detailing, on the right correct detail.

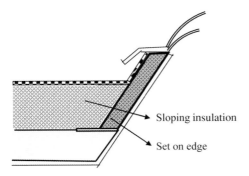

Figure 5.16. Heavy compact roof: dome with correct set on edges.

Transient response

Heavy deck

Starting-point is the assemblies H 1.0, H 2.0 and H 3.0. As a reference, we take more or less the same low-sloped roof example used for the clear thermal transmittance calculations: 14 cm reinforced concrete deck, lightweight concrete sloping screed with a mean thickness of 10 cm, 7 and 17 cm thick dense glass fibre with thermal conductivity of 0.033 W/(m · K). Table 5.4 lists the transient properties.

Table 5.4. Heavy deck, transient properties (daylong period).

U_o (W/(m² · K)) (.¹): real value, insulation thickness)	Variant	Temperature-damping		Dynamic thermal resistance		Admittance	
		D_θ −	ϕ_θ h	D_q m²·K/W	ϕ_q h	Ad W/(m²·K)	ϕ_{Ad} h
0.4 (0.41¹, 7 cm glass fibre)	H 1.0	21.1	13.2	18.4	9.4	1.1	3.8
	H 2.0	78.4	11.7	15.5	10.8	5.1	0.9
	H 3.0	76.9	10.5	15.6	9.7	4.9	0.8
0.2 (0.19¹,17 cm glass fibre)	H 1.0	63.6	17.0	50.1	12.8	1.3	4.2
	H 2.0	217.1	15.2	43.0	14.2	5.1	0.9
	H 3.0	209.6	13.9	42.5	13.1	4.9	0.8

Whether the insulation layer sits under the deck (H 1.0), between deck and sloping screed (H 2.0) or on the sloping screed (H 0.3), even with 7 cm glass fibre the assembly fulfils the requirements $D_\theta \geq 15$ and $D_q \geq 1/U_{max}$ m² · K/W even for $U_{o,max} = 0.1$ W/(m² · K), let it be with 17 cm. The admittance is another story. For H 1.0, insulation under the load bearing deck, the value drops to 1.1–1.3 W/(m² · K), far below the requirement of half the surface film coefficient inside (7.7 W/(m² · K)). H 2.0 and H 3.0 perform excellently, with Ad ≈ 5 W/(m² · K). Or, though the assembly with insulation under the deck still shows quite some thermal inertia, heat storage capacity is neutralized.

5.3 Compact low-sloped roofs

Semi-heavy deck

Assembly SH 1.0 gives the values of Table 5.5. For SH 2.0 (insulation under the deck) and SH 3.0 (insulation on the deck), the table assumes 5 cm PUR as insulation material.

Table 5.5. Semi-heavy deck (cellular concrete), transient properties (daylong period).

Thickness cm	Moisture content	Temperature-damping		Dynamic thermal resistance		Admittance	
		D_θ –	φ_θ H	D_q m²·K/W	φ_q h	Ad W/(m²·K)	φ_{Ad} h
15	Air-dry	3.0	7.1	1.6	4.7	1.9	2.4
	Building moist	4.6	8.4	1.4	6.5	3.4	1.9
20	Air-dry	5.3	9.2	2.6	6.8	2.0	2.4
	Building moist	8.6	10.8	2.6	8.9	3.4	1.9
28	Air-dry	12.8	12.5	6.4	10.2	2.0	2.3
	Building moist	23.5	14.6	7.0	12.8	3.3	1.9
+5 cm PUR below	Air-dry	9.8	12.8	13.8	9.5	0.7	3.3
+5 cm PUR on top	Air-dry	31.5	12.1	16.7	10.0	1.9	2.2

None of the roofs simultaneously fulfil the requirements for temperature damping beyond 15 and admittance above half the surface film coefficient indoors (h_i = 7.7 W/(m²·K)). While for a moist and an air-dry 28 cm with 5 cm PUR on top, temperature damping clearly passes 15, admittance stays below 3.9 W/(m²·K). Neither deck thickness nor insulation on top affects this property. Building moisture does. The reason is simple. Moist cellular concrete has a higher volumetric specific heat capacity (ρc) and a higher thermal conductivity (λ), resulting in a larger contact coefficient ($b = \sqrt{\rho c \lambda}$) and thus a higher admittance. For the temperature damping and dynamic thermal resistance instead, thickness and insulation on top matters. Stepping from 15 to 28 cm for example quadruples both values.

Light-weight deck

With PUR as insulation, 22 mm plywood as boarding and gypsum board as inside finish, one gets the temperature damping, dynamic thermal resistance and admittance values of Table 5.6. The message is clear: low-sloped roofs with light-weight decks do not fulfil the requirements whatever the assembly might be. Yet, L 3.0 (thermal insulation on the boarding) performs better then L 1.0 (thermal insulation under the boarding). Nonetheless, lightweight roofs are not by definition the cause of summer overheating in the spaces below. Whether complaints will surface, depends much more on glass orientation, glass area, glass type, well or no solar shading, ventilation strategy and thermal admittance of partition walls and floor then on the thermal inertia and storage capacity of the roof. Of course, its dynamic thermal resistance must be high enough, which is a problem with none of the variants.

Table 5.6. Light-weight deck, transient response (daylong period).

U_o (W/(m²·K)) (real value, insulation thickness)	Variant	Temperature-damping		Dynamic thermal resistance		Admittance	
		D_θ –	ϕ_θ h	D_q m²·K/W	ϕ_q h	Ad W/(m²·K)	ϕ_{Ad} h
0.4 (0.35, 6 cm PUR)	L 1.0	2.0	5.6	3.0	2.1	0.7	3.5
	L 3.0	5.5	6.6	3.5	3.4	1.5	3.2
0.2 (0.18, 13 cm PUR)	L 1.0	4.3	8.0	6.3	3.6	0.7	4.4
	L 3.0	12.6	8.2	7.7	4.9	1.6	3.3

To conclude, semi-heavy and lightweight roofs show inferior transient response, whereas for heavyweight ones crucial is where the insulation sits. On or under the sloped screed is excellent, under the deck not. Nevertheless overheating risk depends on so many other design decisions, that none of these assemblies should be rejected.

Durability

Although durability depends on more than just hygrothermal loading, this is the aspect discussed here because it helps in establishing the best performing variants.

Heavy deck

We use again the variants H 1.0, H 2.0 and H 3.0:

Assembly H 1.0	Assembly H 2.0	Assembly H 3.0
Roofing membrane	Roofing membrane	Roofing membrane
Sloping screed	Sloping screed	*Thermal insulation*
Deck	*Thermal insulation*	Sloping screed
Thermal insulation	**Deck**	**Deck**
Inside finish	Inside finish	Inside finish

For an assembly with 17 cm mineral wool, Table 5.7 and Figure 5.17 give the temperature of the roofing membrane, the lightweight concrete sloped screed and the reinforced concrete deck during a cold winter and hot summer day in a moderate climate.

With the insulation under the deck – H 1.0 – this and the sloped screed are subjected to large annual temperature differences with quite important thermal movements as a result. As one cannot divide a load bearing deck in smaller lengths along the span, the movements bend the supporting columns and walls. With masonry, this may cause severe cracking. A way out is to limit the annual temperature differences in the deck, in other words, moving the insulation from under to between deck and screed, i.e. H 2.0. That way, the deck's temperature load reduces drastically. The screed however suffers even more. But now dividing is possible. An expansion joint along the roof edges and joints all over the surface, cutting the screed in ±25 m² large fields, suffices. Because this is too often theoretical, a more realistic way to minimize the screed's temperature load is by moving the insulation on top of it – H 3.0 –. As temperature swings in the membrane then increase, classic bitumen felts no longer comply as the roofing's top layer.

5.3 Compact low-sloped roofs

Table 5.7. Heavy deck, temperatures in all layers.

14 cm reinforced concrete deck 10 cm sloped screed 17 cm MW	Temperatures, Cold winter day °C		Temperatures, Hot summer day °C		Annual diff.
	Min.	Max.	Min.	Max.	°C
H 1.0 Insulation below the heavy deck					
Deck	−14.4	−11.9	29.5	36.8	51.2
Sloped screed	−16.9	−8.6	24.7	45.7	62.6
Membrane	−19.9	−3.7	16.9	> 60	> 80
H 2.0 Insulation between heavy deck and screed					
Deck	15.6	15.7	23.3	23.6	8.0
Sloped screed	−18.9	−6.6	17.3	53.5	72.4
Membrane	−19.1	−3.6	14.8	61	> 80
H 3.0 Insulation above sloped screed					
Deck	15.6	15.7	23.3	23.6	8.0
Sloped screed	14.7	15.0	23.4	24.1	8.4
Membrane	−21.3	0.6	12.3	> 70	> 91.3

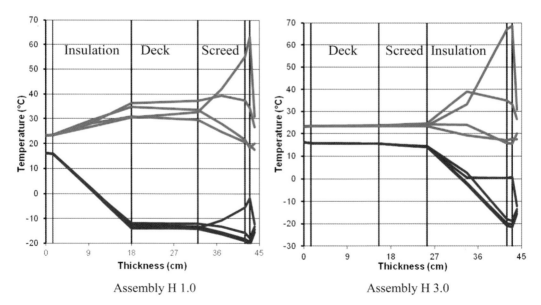

Assembly H 1.0   Assembly H 3.0

Figure 5.17. Low-sloped roof with heavy concrete deck, temperatures during a cold winter (bottom) and hot summer day (top) in a moderate climate.

Hygrically an analogous story holds: larger relative humidity fluctuation in the deck with H 1.0 than with H 2.0 and H 3.0. In other words, the problems assembly H 1.0 may cause in terms of hygrothermal loading look threatening enough to skip that variant. Remain:

Assembly H 2.0	Assembly H 3.0	Assembly H 2.1	Assembly H 3.1	Assembly H 3.2
Roofing membrane	Roofing membrane	Roofing membrane	Roofing membrane	Roofing membrane
Sloping screed	*Thermal insulation*	Sloping screed	*Thermal insulation*	*Thermal insulation*
Thermal insulation	Sloping screed	*Thermal insulation*	Sloping screed	Vapour barrier
Deck	**Deck**	Vapour barrier	Vapour barrier	Sloping screed
Inside finish	Inside finish	**Deck**	**Deck**	**Deck**
		Inside finish	Inside finish	Inside finish

Of course, on site cast concrete decks show shrinkage. The remedy is a well-balanced casting scheme and retarding drying.

Semi-heavy deck

Basic variants still are:

Assembly SH 0.0	Assembly SH 1.0	Assembly SH 3.0
Roofing membrane	Roofing membrane	Roofing membrane
Deck	**Deck**	*Thermal insulation*
(Inside finish)	*Thermal insulation*	**Deck**
	Inside finish	Inside finish

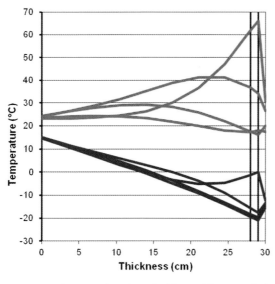

Figure 5.18. Low-sloped roof with semi-heavy cellular concrete deck, assembly SH 0.0, temperatures during a cold winter (bottom) and hot summer day (top) in a moderate climate.

5.3 Compact low-sloped roofs

The cellular concrete in SH 1.0 suffers from large temperature differences with consequences annoying enough to skip the variant. SH 3.0 instead offers the same advantages as H 3.0: the load bearing deck is excellently protected. SH 0.0 remains. Figure 5.18 portrays the temperature during a cold winter- and hot summer day in a 28 cm thick cellular concrete deck. Not only does the mean vary by a total of 37.7 °C, the differences across also fluctuate greatly.

Through it, the deck elements not only expand and contract, they also bend convex in summer and concave in winter with a total displacement at the centre of the span:

$$y = \frac{\alpha \, L^2 \, \Delta\theta}{4 \, d} \tag{5.3}$$

In this equation, α is the thermal expansion coefficient, and d deck thickness and L span. For a 6 meter span, the annual change in length at the supports totals 3.5 mm, whereas bending displacement equals 8.4 mm, 4.6 mm down, and 3.8 mm up. Both have consequences. For a roofing membrane fully bonded to the cellular concrete, tearing probability at the supports is high. This and bending recommend two measures: (1) loosely lay the roofing membrane at all supports over some 25 cm, this at both sides of the contacting deck elements, and (2) allow for 2 cm wide, elastically sealed joints above all non-bearing partition walls touching the deck, (Figure 5.19). With these precautions, assembly SH 0.0 and SH 0.1 succeeds. This leaves:

Assembly SH 0.0	Assembly SH 0.1	Assembly SH 3.0	Assembly SH 3.1	Assembly SH 3.2
Roofing membrane	Roofing membrane	Roofing membrane	Roofing membrane	Roofing membrane
Deck	**Deck**	*Thermal insulation*	*Thermal insulation*	*Thermal insulation*
(Inside finish)	Vapour barrier	**Deck**	**Deck**	Vapour barrier
	Inside finish	Inside finish	Vapour barrier	**Deck**
			Inside finish	Inside finish

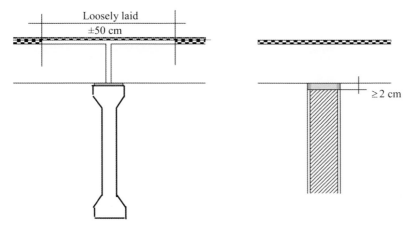

Figure 5.19. Semi-heavy cellular concrete deck, roofing membrane loosely laid above supports, joint between partition walls and deck.

Light-weight deck

Until now, no variants were skipped (included L 1.1 and L 3.1):

Assembly L 1.0	Assembly L 3.0
Roofing membrane	Roofing membrane
Deck	*Thermal insulation*
Thermal insulation	**Deck**
Inside finish	Inside finish

Thermal loading of the deck is highest for L 1.0 and L 1.1. Yet, this should not be a problem. Most probably, particle, plywood or OSB boards with limited dimensions (2.4 × 1.2 m) are used as deck material, all three being wood-based composites with low thermal expansion coefficient. However, things can go wrong hygrically, which is why we eliminate L 1.0. This leaves:

Assembly L 3.0	Assembly L 3.1
Roofing membrane	Roofing membrane
Thermal insulation	*Thermal insulation*
Deck	Vapour retarder
Inside finish	**Deck**
	Inside finish

The conclusion is clear: the heavy, semi-heavy or lightweight deck assemblies with the insulation directly under the roofing membrane perform the best, although the membrane must then withstand higher thermal loads, a fact excluding bituminous roofing felts as a top layer.

Moisture tolerance

Rain, building moisture, and interstitial condensation play the main role. Of course, relative humidity at ceiling level exceeding 80% on a monthly basis and surface condensation may also play a role. But with the clear thermal transmittances imposed today, the only reason for this are possible thermal bridges or insufficient ventilation of the spaces below.

Rain

Roofing membranes figure as one-step rain screens. Once leaky, rainwater-soaking starts. In an assembly containing no or a perforated vapour barrier, or, a vapour barrier at the wrong location, all capillary materials present will suck rainwater. Afterwards evaporation and diffusion may humidify the thermal insulation, even when not capillary and quite vapour retarding (only cellular glass will stay dry, at least if not freezing). Ultimately, the whole roof section can turn saturated, which is detrimental from a durability point of view. Capillary materials turn wet above capillary. Thermal conductivity increases. Wet mineral wool boards lose strength and stiffness. Bar corrosion in concrete accelerates. Frost weathers cellular concrete. Wood-based board materials start rotting, the roof drips. Rain leakage must thus be avoided. The best guarantee is a correctly assembled low-sloped roof, high quality properly laid roofing membranes, compartmentalisation of the thermal insulation to restrict the leakage spread, correct detailing, etc. (see design and execution).

5.3 Compact low-sloped roofs

After repair of a leaky roofing membrane, the following situations are possible:

- *No vapour barrier*
  In this case, all materials around the leak will be wet, which is comparable with building moisture in a roof without vapour barrier
- *Tight vapour barrier*
  This prevents the leaking water from reaching the layers below. Of course water will pond on it. This is comparable to a roof water-tightened during rainy weather, just like building moisture again.
- *Leaky vapour barrier*
  Combines the two cases above

Or, to know if the assembly will dry, find out what happens with building moisture.

Building moisture

Heavy deck

Not only the deck and the screed but sometimes also the insulation may contain building moisture. The quantities can be impressive, as controls on a couple of roofs, just before waterproofing, showed, see Table 5.8.

Table 5.8. Building moisture in heavy deck low-sloped roofs, measured on site.

Layer	Building moisture kg/m³	Cap. moisture content kg/m³
Sloped screed	38–366	105
Levelling layer sloped screed	154	
Thermal insulation, expanded perlite boards	41	70
Thermal insulation, dense mineral wool boards	45	–

A sloped screed is best considered capillary wet. The same applies for the heavy deck. Insulation wetness instead often follows from condensation of building moisture present in the two layers. What happens after waterproofing depends on the assembly:

Assembly H 2.0	Assembly H 3.0	Assembly H 2.1	Assembly H 3.1	Assembly H 3.2
Roofing membrane	Roofing membrane	Roofing membrane	Roofing membrane	Roofing membrane
Sloping screed	*Thermal insulation*	Sloping screed	*Thermal insulation*	*Thermal insulation*
Thermal insulation	Sloping screed	*Thermal insulation*	Sloping screed	Vapour barrier
Deck	Deck	Vapour barrier	Vapour barrier	Sloping screed
Inside finish	Inside finish	Deck	Deck	Deck
		Inside finish	Inside finish	Inside finish

The deck of H 2.0 dries to the inside. In winter, building moisture from the deck also diffuses across the insulation into the sloped screed where it condenses. During warm weather, diffusion moves the other way, back to the deck (Figure 5.20). How wet the insulation becomes, depends on its vapour resistance factor, see Figure 5.21.

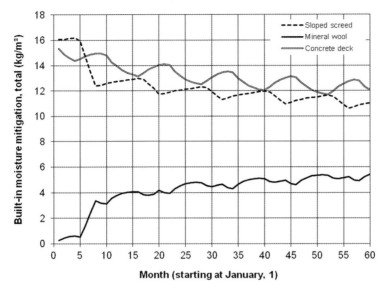

Figure 5.20. Assembly H 2.0 insulated with 10 cm thick, dense mineral wool boards. Building moisture from deck and sloped screed condensing in the insulation, deck and sloped screed drying slowly (calculated with Match©).

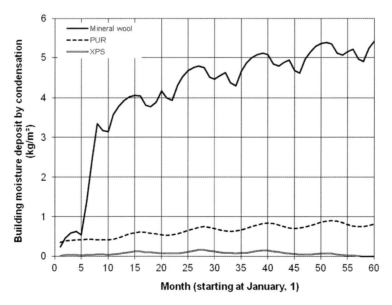

Figure 5.21. Assembly H 2.0. Building moisture from deck and sloped screed condensing in mineral wool ($\mu = 1.2$), PUR ($\mu = 30$) and XPS ($\mu = 120$) insulation (calculated with Match©).

5.3 Compact low-sloped roofs

The higher the vapour resistance factor is, the lower the amounts condensing but the slower the screed dries, with all the risks this implies: weathering by frost, faster degradation of the membrane, etc. With very vapour permeable insulating materials, the reversed vapour flow can create water ponds on a less capillary deck, with water dripping into the ceiling luminaries when the guide rods lie on and perforate the deck. Even with the deck air-dry, screed and insulation may still stay wet for so many years that H 2.0 is best skipped.

H 2.1? This variant also brings no relief. Thanks to the vapour barrier, diffusion from the deck to the sloped screed stops, but each summer building moisture from the screed again condenses in the insulation and on the vapour barrier. In early winter, that wetness diffuses back to the screed where it condenses with latent heat released temporarily increasing heat loss. Besides, the screed will never dry, with all risks mentioned. Also H 2.1 is best omitted.

The H 3 series remains:

Assembly H 3.0	Assembly H 3.1	Assembly H 3.2
Roofing membrane	Roofing membrane	Roofing membrane
Thermal insulation	**Thermal insulation**	**Thermal insulation**
Sloping screed	Sloping screed	<u>Vapour barrier</u>
Deck	<u>Vapour barrier</u>	Sloping screed
Inside finish	**Deck**	**Deck**
	Inside finish	Inside finish

H 3.0 and H 3.1 both perform badly. Although the screed in H 3.0 will dry faster than in H 2.0 and H 2.1, due to building moisture diffusing into the insulation where it condenses, moisture content there may increase quickly during the first years. A vapour permeable insulation material such as mineral wool can collect 14 kg of water per m^2 after five years, increasing the clear thermal transmittance with 30 à 40% (Figure 5.22), while the boards lose stiffness.

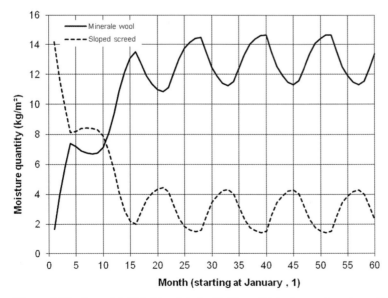

Figure 5.22. Assembly H 3.0 insulated with 10 cm thick, dense mineral wool boards. Building moisture from sloped screed condensing in the insulation, sloped screed dries (calculated with Match$^©$).

Vapour retarding insulation limits the problems. The amounts condensing in fact are inversely proportional to the vapour resistance factor, albeit synthetic foams will collect enough moisture to show irreversible swelling under solar radiation. Only cellular glass remains dry. In the long run, even mineral wool boards will dry but this takes years even with a vapour permeable deck. What about a non-capillary, vapour retarding deck? Dryness then shifts unacceptably far into the future. Z 3.1 moves that moment still farther away. The vapour barrier prevents the screed from drying while the insulation can become wet. Therefore, neither H 3.0 nor H 3.1 should be applied.

We are thus left with H 3.2. With a perfectly fixed vapour barrier of correct quality, this assembly excludes building moisture from condensing in the insulation. Through that, the likelihood of durability problems is extremely small. Or, H 3.2 is the correct choice, top-down:

Roofing membrane
Thermal insulation
Vapour barrier
Sloping screed
Deck
Inside finish

Semi-heavy deck

Fresh cellular concrete contains lots of building moisture. A low-sloped roof with such a deck must thus have the possibility to dry damage-free. Yet, five variants are still competitive:

Assembly SH 0.0	**Assembly SH 0.1**	**Assembly SH 3.0**	**Assembly SH 3.1**	**Assembly SH 3.2**
Roofing membrane	Roofing membrane	Roofing membrane	Roofing membrane	Roofing membrane
Deck	**Deck**	*Thermal insulation*	*Thermal insulation*	*Thermal insulation*
(Inside finish)	Vapour barrier	**Deck**	**Deck**	Vapour barrier
	Inside finish	Inside finish	Vapour barrier	**Deck**
			Inside finish	Inside finish

SH 0.1 may be omitted. Building moisture in fact remains locked between two vapour tight layers, preventing the cellular concrete from drying while keeping its thermal conductivity above 0.29 W/(m · K), i.e. more than two times the air-dry value. In addition, the likelihood to get blisters in the membrane and vapour retarder is too high. SH 3.0 and SH 3.1 face the same problems as the heavy deck variants H 3.0 and H 3.1: condensation of building moisture in the thermal insulation. If for variant SH 3.0 the reversal from condensation to drying takes a couple of years, for variant SH 3.1 ten years and more are needed. Meanwhile the membrane may catch irreversible damage. Or, only SH 0.0 and SH 3.2 withstand the building moisture check:

Assembly SH 0.0	**Assembly SH 3.2**
Roofing membrane	Roofing membrane
Deck	*Thermal insulation*
(Inside finish)	Vapour barrier
	Deck
	Inside finish

5.3 Compact low-sloped roofs

Light-weight deck

With lightweight decks, the situation is less complicated. Structural sheet-metal plates cannot contain building moisture and timber-based materials require air-dryness when applied. Nevertheless, some caution must prevail. With structural sheet-metal plates, rainwater may collect in the valleys during construction. Timber-based materials in turn are often processed while too humid. For those reasons, variant L 3.1 is preferred, although L 3.0 may also perform well. No vapour retarder in fact turns this variant into a self-drying assembly.

Assembly L 3.0	Assembly L 3.1
Roofing membrane	Roofing membrane
Thermal insulation	***Thermal insulation***
Deck	Vapour retarder
Inside finish	**Deck**
	Inside finish

Interstitial condensation

For a long time, interstitial condensation in the air-dry roof was perceived as the big culprit. The reason is given above, when discussing typology. Rain proofing low-sloped roofs demands a watertight layer at the outside. Because such layers are also vapour tight, we get an assembly with the outside layer acting as vapour barrier. A Glaser diagram then shows this unavoidably results in condensation underneath (Figure 5.23). But the condensing quantities and the balance between winter humidification and summer drying are seldom calculated.

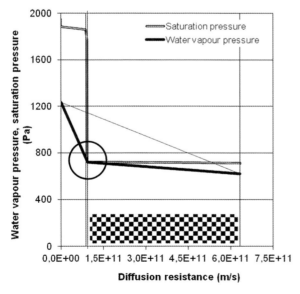

Figure 5.23. Assembly H 3.0, the Glaser diagram indicates condensate depositing underneath the membrane.

Heavy deck

Assume the roof is airtight. With heavy deck roofs, annually cumulating condensate is excluded in indoor climate class 1 to 3. Figure 5.24 shows the maximum deposited in 3 after a moderate climate winter, depending on the diffusion resistance of the assembly without membrane. Only very vapour permeable decks look problematic. Anyhow, from indoor climate class 4 on, deposit accumulates over the years, slowly wetting the insulation, see Figure 5.25. Avoidance requires a vapour retarding layer directly under that layer.

Let it be noted that in the calculation model used, keeping the vapour pressure in the hygric centre of gravity of the assembly at the annual mean indoors introduced hygric inertia, location of that centre being given by:

$$x_{RH} = \frac{\sum_{n=1}^{2}\left(a_{Hj}\, d_j\, x_j\right)}{\sum_{n=1}^{2}\left(a_{Hj}\, d_j\right)} \tag{5.4}$$

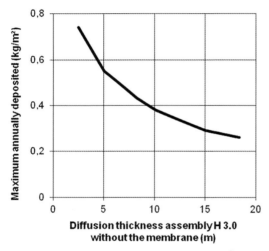

Figure 5.24. Assembly H 3.0, $U = 0.2$ W/(m$^2 \cdot$ K), indoor climate class 3, maximum deposit underneath the membrane after the winter.

Figure 5.25. Heavy deck roof without vapour retarder above an indoor climate class 5 building, interstitial condensation in the PUR-insulation.

5.3 Compact low-sloped roofs

with $a_{H,j}$ the hygroscopic constant and d_j the thickness of both the deck and the screed. x_j is the distance between the centre of both layers and the assembly's underside. Due to inertia, more condensate was deposited than predicted by a common Glaser calculation. In spite of that, the annual indoor climate class 3 maximum remains far below 1 kg/m², even for roof assemblies with limited diffusion thickness without membrane. This 1 kg/m² is surely not an absolute upper limit value. If for example the requirement should be that after each winter the clear thermal transmittance should not show more than a 5% increase, then for 17 cm of insulation the deposit might touch 2.2 kg/m. Nevertheless, we fix the limit at 1 kg/m². In fact, during sunny days in springtime, vapour flow goes the other way around, from the insulation to the screed, to turn back direction insulation at night, with companion latent heat flow increasing the clear thermal transmittance. This is considered unwanted.

Semi-heavy deck

The same conclusions hold. In the indoor climate classes 4 and 5, annually accumulating condensate is a fact and a vapour retarder below the deck a necessity. Building moisture, however, excludes this solution. SH 3.2 therefore is the designated choice here.

Light-weight deck

The first concern here is air-tightness. If guaranteed, then Figure 5.24 also holds for light-weight decks: the lower the diffusion resistance of the assembly without membrane, the higher the annual condensation maximum deposited underneath. The acceptable maximum again is 1 kg/m². An additional argument to keep that number is the lower moisture tolerance of most light-weight decks compared to concrete, with increased damage risk when the condensate underneath the membrane repeatedly evaporates during sunny days to diffuse across the insulation to condense on the deck. With organic insulation materials, the maximum allowed is 3% kg/kg, a value that makes a vapour retarder indispensible whatever the circumstance may be.

Conclusion

Do compact roofs need a vapour retarder? In moderate climates, building moisture and interstitial condensation shaped the decision array of Table 5.9 (–: no vapour retarder; E1, E2, vapour retarder of that class).

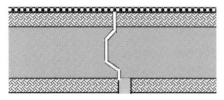

Figure 5.26. Sandwich elements, critical are the joints between elements.

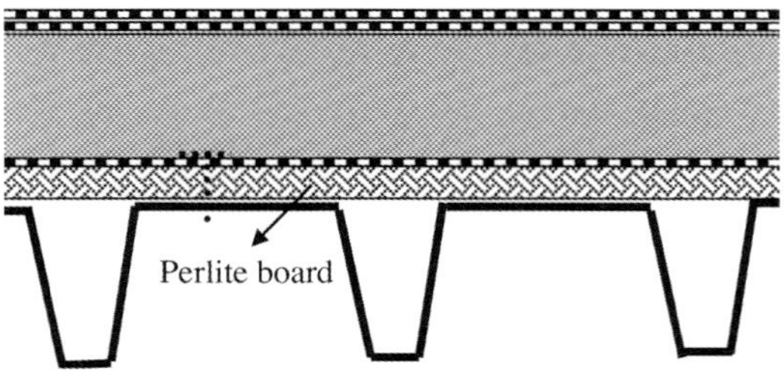

Figure 5.27. Sheet-metal deck above buildings ICC 4 and 5.

Table 5.9. Low-sloped roof, vapour retarder?

Heavy deck, sloping screed in light-weight concrete						
ICC	Minimal vapour retarder/barrier quality					
	Cork	MW	Cell. Glass	PUR	Phenol	Perlite
1, 2, 3	E2	E2	(E2)[2]	E2	E2	E2
4, 5	E4	E4	(E2)[2]	E4	E4	E4
Heavy deck, no sloping screed, sloping screed in building moisture poor no-fines concrete						
ICC	Minimal vapour retarder/barrier quality					
	Cork	MW	Cell. Glass	PUR	Phenol	Perlite
1, 2, 3	E2	E2	–	–	–	E2
4, 5	E3	E3	(E2)[2]	E3	E3	E3
Semi-heavy deck, assembly SH 0.0						
ICC	Vapour retarder?					
1, 2, 3	Unwanted					
4, 5	Assembly unfit					
Semi-heavy deck, assembly SH 3.1						
ICC	Minimal vapour retarder/barrier quality					
	Cork	MW	Cell. Glass	PUR	Phenol	Perlite
1, 2, 3	E2	E2	(E2)[2]	E2	E2	E2
4, 5	E3	E3	(E2)[2]	E3	E3	E3
Light-weight deck, wood or wood-based						
ICC	Minimal vapour retarder/barrier quality					
	Cork	MW	Cell. Glass	PUR	Phenol	Perlite
1	–[1]	–[1]	–[1]	–[1]	–[1]	–[1]
2, 3	E2	E2	–[1]	E2	E2	E2
4, 5	E3	E3	(E2)[2]	E3	E3	E3
Light-weight, steel deck						
ICC	Minimal vapour retarder/barrier quality					
	Cork	MW	Cell. Glass	PUR	Phenol	Perlite
1	–	–	–[2]	–	–	–
2, 3	E1	E1	–[2]	–[3]	–[3]	E1
4, 5	E3[4]	E3[4]	(E2)	E3[4]	E3[4]	E3[4]
Light-weight deck, sandwich elements (Figure 5.26)						
ICC	Vapour retarder/barrier?					
1, 2, 3	Superfluous on condition the elements have a diffusion thickness equal to a vapour retarder E1					
4. 5	E3					

5.3 Compact low-sloped roofs

Legend for Table 5.9:
1. Before mounting the insulation, seal all joints between deck elements with glass fibre bitumen.
2. Cellular glass is a vapour tight insulation material. If perfectly mounted (i.e. boards pressed in bitumen, insulation finished with bitumen), a vapour retarder is superfluous. Because in building practice perfection is never guaranteed, mounting a vapour retarder first is an advisable redundant measure.
3. On condition the boards are lined with glass fibre bitumen. If not, a vapour retarder E1 is needed.
4. The vapour barrier must be mounted on a substrate, for example perlite boards. The two layer barrier comes on top, with the first layer fixed mechanically across the board into the deck (Figure 5.27).

Thermal bridges

See 'thermal transmittance'.

5.3.2.3 Building physics: acoustics

With heavy deck sound insulation is not a problem. Mass takes cares of it. For a 14 cm or thicker concrete floor, average sound transmission loss passes 72 dB. Things differ for semi-heavy and lightweight decks. A 10 cm thick cellular concrete SH 0.0 assembly only gives 37 dB. For 20 cm, this is 43 dB. Doing better demands composite assemblies, which is indisputably the case for lightweight decks.

5.3.2.4 Fire safety

The roof of medium and high rise buildings figures as evacuation platform. That is why the same requirements hold as for partitions between fire compartments: a 90′ fire resistance (structural integrity, temperature increase, and smoke tightness). Moreover, membrane and insulation must be hardly flammable and covered in a way burning is unlikely, which demands ballast or a green roof finish. For low-rise premises, requirements are less strict.

5.3.2.5 Maintenance

A long service life presumes regular maintenance. In moderate and cold climates, one must remove leaves, moss, plants, and other vegetation before each winter. Afterwards, a general inspection with control of all junctions (flashings, edges, etc.) is recommended. Where needed, membranes are repaired. Downpipes must stay functioning and trafficked zones demand a more than annual control (for example the route between roof entrance and fans, cooling towers and others).

5.3.3 Design and execution

5.3.3.1 Assembly

The checks showed one assembly was the best for heavy deck roofs:

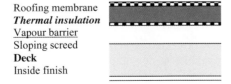

Roofing membrane
Thermal insulation
<u>Vapour barrier</u>
Sloping screed
Deck
Inside finish

Deck and sloping screed belong to the carcass work. The roofing contractor layers the vapour barrier, the insulation and the roofing membrane, while the inside finish is part of the completion work. Polymer bitumen mounted as a roofing membrane is preferred as a vapour retarder. Vapour tightness in fact presumes waterproofing. A PE-foil perhaps has the correct vapour resistance but is too perforation sensitive and folds when walked upon. It is also blown easily by wind and does not allow gluing the package vapour retarder/thermal insulation/membrane at the sloping screed. For those reasons, there is no guarantee of vapour tightness.

Roofs with semi-heavy and light-weight deck do not need a sloped screed if the deck gets a slight slope. With lightweight deck roofs, the down-up layer sequence at the top remains: vapour retarding layer (if needed), thermal insulation, and roofing membrane. When the thermal transmittance requirements allow, semi-heavy deck roofs above indoor climate class 1, 2 and 3 buildings do not need vapour retarders and thermal insulation. Above indoor climate class 4 and 5 buildings the assembly does not differ from lightweight deck roofs. Yet, one weakness remains: when rainwater gets trapped during construction between vapour retarder and waterproof membrane, it cannot dry. This increases the clear thermal transmittance and may blister the membrane and degrade the insulation. A possibility, which of course is not absolutely effective, consists of compartmentalizing the insulation by closing it up at regular distances, using a watertight strip that couples the vapour retarder to the membrane (Figure 5.28).

Figure 5.28. Closing up the thermal insulation.

Some manufacturers and contractors offer alternatives. One variant combines sloped screed and insulation by embedding EPS-blocks in EPS-concrete. No problems occurred. In fact, the high vapour permeability of the composite screed curbs local high vapour pressures, eliminating the likelihood of blistering risk and minimizing visible damage. However, in the first years, the moist EPS-concrete wets the EPS-blocks, followed later by slow drying to the inside. During this period, thermal transmittance is higher then predicted assuming air-dry materials. The solution thus performs worse than the best choice.

5.3.3.2 Details

Sloped screed

Zero-sloped roofs increase leakage risk. Therefore, a slight slope of 1.5 cm/m is preferred. Sloping may be done in one or two directions. The first creates an ensemble of flat surfaces, the second an involution plane, which makes gutter channels superfluous but causes problems with stiff insulation boards. These are hardly adaptable to such a plane.

Sewers and downpipes

The roof should drain rain via well adapted leaf-basket protected sewers, not as shown in Figure 5.29, into downpipes. Although not a favourite of architects, downpipes are best situated outside the enclosure. That way, possible leakage does not end in annoying moisture damage.

5.3 Compact low-sloped roofs

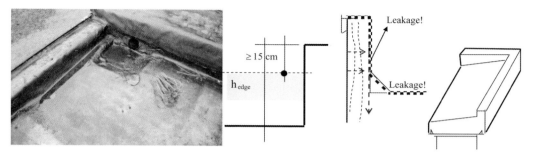

Figure 5.29. Left: sewer, badly detailed leaf basket removed.
Right: roof edges with height, water proofing, cover stones.

Roof edges

Roof edge height, measured from the top of the load bearing deck, must obey following inequality (all measures in cm but L in m):

$$h_{edge} \geq 7 + 1.5\,L + d_{ins} + 15 \tag{5.5}$$

with L half the distance between two downpipes along the edge, d_{ins} the insulation thickness in cm and 15 (cm) minimum height above the roofing membrane (Figure 5.29). Roof edges are water tightened over their height at the roof-side and on top, using additional membranes, which are glued or burned to the roofing membrane. At the façade side, they get a roof edge trim, detailed in a way thermal expansion and shrinkage is possible without tearing the membrane. Aluminium roof edgings therefore require limited lengths (±3 m) with sliding couplers in between. An alternative are cover stones (Figure 5.29). These should slope towards the roof surface, while a small fold up and little overhang with drop in front must prevent rain from reaching the vulnerable joint between edge and membrane. Edges may also not facilitate rain seepage to the sloping screed and the load bearing deck. Many designers do not like roof overhangs, despite the many advantages they offer. They protect the upper part of the façade against wind-driven rain and staining. A thermal bridge free design of course is not simple. Finally, keeping the roof edges top sides in a same horizontal plane makes water proofing easier. Avoid sudden steps in edge height!

Roof building-ups

Raising the membrane against the wall 10 cm above the roof edges guarantees water proofing at roof building ups. A cover flashing, embedded in the building up walls acts as protection. At cavity walls, these flashings are fixed in the veneer below the cavity tray. The 10 cm above the roof prevents rainwater in case of roof flooding from entering the building before flowing over the edges (Figure 5.30).

Figure 5.30. Flashing.

Settlement and expansion joints

The roofing membrane should have the opportunity to expand and contract without either cracking or having settlement and expansion joints that collect water. Therefore, they are raised and finished up to the height of the edges. Figure 5.31 gives an example.

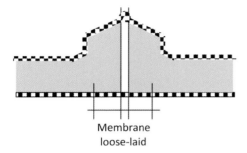

Figure 5.31. Expansion joint.

5.3.3.3 Special low-sloped roof uses

Green roofs

Green roofs became popular with the increased interest in sustainable construction. Soil and greenery help in buffering rainwater. Widespread use in urban environments may moderate the heat island effect, while the application is quite effective in dampening heat gain indoors during hot weather, mainly due to evaporative cooling. However, contrary to what hardcore believers claim, measurements showed negligible energy benefits during the heating season. The humid soil in fact hardly adds insulation value, so that green roofs do not allow economizing on insulation thickness.

A correct assembly is of true importance. From the bottom upwards we must have the following layers:

1. *High quality, root resistant roofing membrane*
   Minimal leakage risk is imperative. In fact, finding leaks is a disaster as greenery, soil, drainage layer, all have to be removed! Root resistivity prevents leakage by avoiding roots from perforating the membrane.
2. *Separating layer*
   Allows some movement between green roof package and roofing membrane
3. *Drainage layer*
   Enables correct dewatering of the soil
4. *Filter textile*
   Avoids that soil particles seep the drainage layer
5. *Soil*
   Thickness depends on the kind of greenery

The extra load exerted by the green roof has to be taken into account during design.

5.4 Protected membrane roofs

Parking roofs

The biggest problem is withstanding the horizontal braking forces, which pull together the roofing membrane. Execution of that membrane, as a single mutually fully adhering package with the insulation and the vapour retarder, bonded to the sloping screed with hot bitumen, minimizes the problem. The insulation must have high compression and fair shear resistance. Cellular glass meets that requirement. The parking deck itself consists of a poured mastic asphalt floor or of large concrete plates, which load the membrane via support strips, bedplates, or a gravel filter. This allows dewatering underneath.

Roof terraces

The vertical loads on roof terraces are much lower than those on parking roofs, while horizontal forces hardly intervene. Including the membrane, the assembly rules for compact low-sloped roofs apply. The membrane anyhow gets a walked on finish. Best choice is heavy terrace tiles on stilts. Hardwood lattice tiles are an alternative. Tiles in mortar bed are not advised. The mortar usually gets soaked with water, which often ends in severe frost damage to the tiled floor. Shrinkage and hygrothermal movements of the mortar could also crack and fold the membrane.

5.4 Protected membrane roofs

5.4.1 In general

The insulation layer and roofing membrane change position in protected membrane roofs. This is only possible with an insulation material, which neither sucks water nor wets under water heads. Two insulation materials apparently meet this criterion: XPS and cellular glass. However, surface wet cellular glass progressively crumbles when freezing, which is why only XPS was used (Figure 5.32). The boards should also not blow away under windy load. The XPS-boards therefore need a heavy gravel or concrete tile ballast.

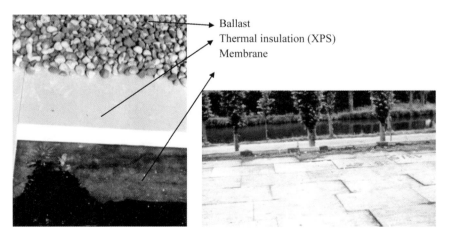

Figure 5.32. Protected membrane roofs: to the left with XPS, to the right with dense mineral wool boards.

Testing in the 1980s also looked at strongly water repellent mineral wool boards. They remained stable without ballast, except at the roof edges and around building-ups. Also the results were hygrically promising but too much erosion finally killed its application.

In terms of performance evaluation, two questions remain: (1) does a protected membrane roof with the same assembly and insulation thickness as a compact roof have an identical whole thermal transmittance and (2) can the insulation layer, even when not capillary and immune to water heads, nevertheless turn wet?

5.4.2 Performance evaluation

5.4.2.1 Thermal transmittance

For protected membrane roofs without filter fabric between ballast and insulation, the answer in heating climates to the first question is negative. Three facts heighten the thermal transmittance: (1) rainwater partially drains underneath the insulation layer, cooling the membrane and increasing heat loss that way; (2) part of this water puddles on the membrane and evaporates across the insulation, with the latent heat coming from indoors; (3) the joints between the insulation boards induce more thermal bridging than is the case for compact roofs. Each needs a correction: ΔU_1 for the drainage underneath, ΔU_2 for the evaporation of water ponds and ΔU_3 for thermal bridging. The effective thermal transmittance then becomes:

$$U_{eq} = U_o + \Delta U_1 + \Delta U_2 + \Delta U_3 \tag{5.6}$$

Correction ΔU_1

Assume the seeping rainwater does not completely fill the air layer between insulation and roofing membrane. The temperature the rainwater reaches there is given by (Figure 5.33):

$$\theta = \theta_\infty + (\theta_o - \theta_\infty)(1+x)^{-\frac{R_1 + R_2 + 4187 R_1 R_2 g_R}{4187 R_1 R_2 g_R}} \tag{5.7}$$

with x the position along the slope, g_R the rain intensity in kg/s, θ_o rain inflow temperature, θ_∞ the final temperature the water may attain, R_1 the thermal resistance between roofing membrane and outdoors and R_2 the thermal resistance between roofing membrane and indoors (both in $m^2 \cdot K/W$). The final value θ_∞ calculates as:

$$\theta_\infty = \frac{R_1 \theta_i + R_2 \theta_e + 4187 R_1 R_2 g_R \theta_o}{R_1 + R_2 + 4187 R_1 R_2 g_R} \tag{5.8}$$

where θ_i is the inside and θ_e the outside temperature. Thermal transmittance during rain increases to:

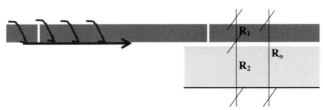

Figure 5.33. Seeping rain.

5.4 Protected membrane roofs

$$U_{pr} = \frac{\theta_i - \theta_\infty}{R_2 (\theta_i - \theta_e)} \tag{5.9}$$

This result is a little too favourable. To what extent, depends on rain intensity and roof dimensions. If we assume the rain is at outside temperature, if for the rain intensity the mean over all rain periods during the heating season pertains and if a fraction f reaches the membrane, then combining (5.8) and (5.9) gives:

$$U_{pr} = \frac{1 + 4187\, R_1\, f\, g_R}{R_1 + R_2 + 4187\, R_1\, R_2\, f\, g_R} \tag{5.10}$$

In case q is the time fraction it rains during the heating season, then the effective thermal transmittance of an inverted membrane roof without filter fabric under the ballast becomes $U_{eq} = (1-q)U_o + q U_{pr}$, with U_o the thermal transmittance the roof should have if compact Correction ΔU_1 then looks like:

$$\Delta U_1 = q\left(U_{pr} - U_o\right) = \left(\frac{4187\, q\, f\, g_R}{1 + 4187\, f\, g_R\, \frac{R_1 R_2}{R_0}}\right)\left(\frac{R_1}{R_0}\right)^2 = \beta \left(\frac{R_1}{R_0}\right)^2 \tag{5.11}$$

The correction increases with (1) higher mean rain intensity (g_R larger) and (2) more often rain during the heating season (q higher), (3) thicker insulation (R_1 higher) and (4) lower thermal resistance of the roof without insulation ($R_0 - R_1$ smaller). If instead f decreases, meaning more rain drains above the insulation, the correction nears zero, reason why today a vapour permeable but quite water tight filter fabric separates the ballast from the insulation.

Table 5.9 and Figure 5.34 show the correction ΔU_1 as a function of period q and rain fraction f. Clearly, protected membrane roofs without filter fabric below the ballast must be discouraged in rainy climates. In fact, the quest for excellent insulation quality looks square with such roof. However, filter fabrics draining 90% of the rain push the correction below 0.01 W/(m² · K), which makes protected membrane roofs a true alternative.

Table 5.9. Correction ΔU_1 for a protected membrane roof with a compact roof thermal transmittance 0.3 W/(m² · K) ($g_R = 3.4 \cdot 10^{-4}$ kg/(m² · s) and $q = 0.077$).

R_1/R_0	ΔU_1 W/(m² · K)		
	$f = 1$	$f = 0.5$	$f = 0.1$
0.9	0.062	0.037	0.009
0.8	0.040	0.025	0.007
0.7	0.027	0.018	0.005
0.6	0.018	0.013	0.004
0.5	0.013	0.009	0.002
0.4	0.008	0.006	0.002

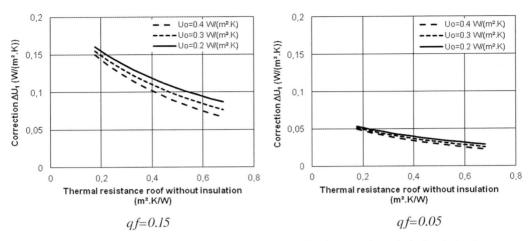

Figure 5.34. Protected membrane roofs, correction ΔU_1 if rain seeps through the insulation layer ($g_R = 3.4 \cdot 10^{-4}$ kg/(m$^2 \cdot$ s)).

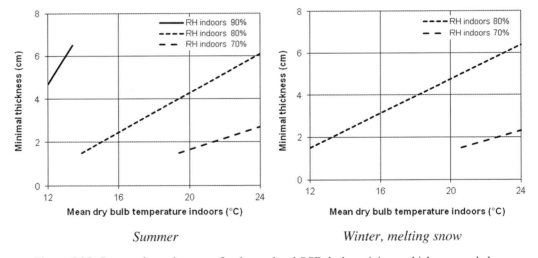

Summary *Winter, melting snow*

Figure 5.35. Protected membrane roofs, plywood and OSB decks: minimum thickness needed.

Seeping through has also other drawbacks. A light-weight deck may cool down so strongly that surface condensation deposits at its inside surface. For that reason the thickness rules of Figure 5.35 must be respected for timber based decks. Corrugated sheet-metal decks are even excluded. Also these restrictions fade away with a filter fabric below the ballast.

Correction ΔU_2

In case seeping causes water ponds under the insulation, vapour pressure there will equal saturation ($p_{sat,c}$). If above vapour pressure outdoors (p_e), evaporating stagnant water will diffuse across the insulation layer to the outside:

5.4 Protected membrane roofs

$$g_v = \frac{p_{sat,c} - p_e}{\dfrac{1}{\beta_e} + \mu N d_{is}}$$ (5.12)

The heat of evaporation comes from inside, giving as steady state heat balance below the insulation:

$$\frac{\theta_i - \theta_c}{R_2} = g_v \, l_b + \frac{\theta_c - \theta_e}{R_1} = U_o + \Delta U_2 \left(\theta_i - \theta_e\right)$$ (5.13)

Correction ΔU_2 then becomes:

$$\Delta U_2 = \frac{U_o}{\theta_i - \theta_e} \left[\frac{R_0 \left(\theta_i - \theta_c\right)}{R_0 - R_1} - 1 \right]$$ (5.14)

In the formulas, l_b is heat of evaporation, μ vapour resistance factor of the insulation layer, g_v the vapour flow rate, R_0 the thermal resistance environment to environment of the roof if compact, R_1 the vapour resistance of the insulation included the surface film resistance outdoors and θ_c the equilibrium temperature of the water below the insulation.

Equation (5.12) shows evaporation and thus heat loss both increase with lower vapour resistance of the insulation. Equation (5.13) corrects that picture a little as more evaporation means lower temperature θ_c and lower vapour saturation pressure below the insulation. But, as Table 5.10 shows, correction for a vapour permeable insulation like mineral wool nevertheless remains so high that only vapour retarding insulation materials as XPS or insulation materials with a vapour retarding underside (the definitive mineral wool boards were manufactured with a bitumen covered underside) should be used. Even with XPS, the boards must fit very well together to keep the impact of the joints on the vapour resistance as low as possible.

Again, a filter fabric under the ballast minimizes puddling and correction ΔU_2.

Table 5.10. Correction ΔU_2 for a protected membrane roof with compact roof thermal transmittance 0.32 W/(m² · K) (outside climate for Uccle).

Month	Correction ΔU_2 W/(m² · K)	
	Mineral wool without bitumenous underside	XPS, closed joints
J	0.14	0.02
F	0.13	0.01
M	0.11	–
A	0.07	–
O	0.07	–
N	0.11	–
D	0.13	0.01

Correction ΔU_3

That correction is given by:

$$\Delta U_3 = \psi_j L_j \tag{5.15}$$

with ψ_j the linear thermal transmittance per meter joint between insulation boards and L the joint length per square meter of roof. The linear thermal transmittance follows from calculation or measurement. For a 5 and 12 cm thick insulation, calculation gave the results of Table 5.11. The table also gives a few measured data. Limiting joint inpact demands large insulation boards. Anyhow, a filter fabric below the ballast is positive as it excludes air washing.

Table 5.11. Joints: ψ-values.

Insulation thickness m	ψ_j, calculated W/(m·K)	
	No air washing	Air washing
0.05	0.016	0.12
0.12	0.003	0.01
Protected membrane roof insulation		Measured ψ_j W/(m²·K)
Mineral wool boards, $d = 0.1$ m		0.0022
Mineral wool boards, $d = 0.1$ m, with bitumen covered underside • No stagnant water below • Stagnant water below		0.00022 0.0022

5.4.2.2 Moisture tolerance

Water ponds below the insulation not only negatively impacts thermal transmittance, it also induces interstitial condensation in the insulation. Indeed, curvature of vapour saturation pressure versus temperature forces vapour pressure to stay saturated until the point of contact for the tangent coming from the vapour pressure outdoors (Figure 5.36).

Condensation deposit per square meter then becomes:

$$m_c = \frac{1}{\mu_{is} N} \left[\left(\frac{dp_{sat}}{dx}\right)_{x=d} - \left(\frac{dp_{sat}}{dx}\right)_{x=\text{contact point } p_e} \right] \Delta t \tag{2.15}$$

In this equation, the derivative for $x = d$ represents the slope of the saturation line in the contact between insulation and puddling water, whereas the derivative in 'x = contact point p_e' gives the slope of the tangent between the saturation line and p_e. The amounts condensing mainly depend on the insulation's vapour resistance factor. If high, hardly some will deposit. If low, quantities may be quite high. In mineral wool boards without a bitumen lined underside, some 8.3 kg/m² will condense during a moderate climate winter. XPS instead will see some 0.08 kg/m². Yet, according to Figure 5.37, a bitumen lined underside excludes interstitial condensation, even with mineral wool.

5.4 Protected membrane roofs

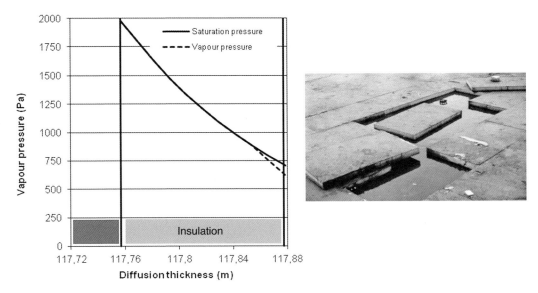

Figure 5.36. Protected membrane roof with stagnant water below the insulation. Mean vapour pressure in the insulation during the month of January (moderate climate of Uccle).

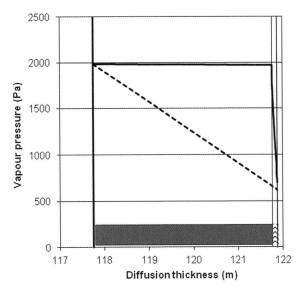

Figure 5.37. Protected membrane roof with stagnant water below the insulation. Mean vapour pressure in mineral wool boards with bitumen at the underside (January, moderate climate of Uccle).

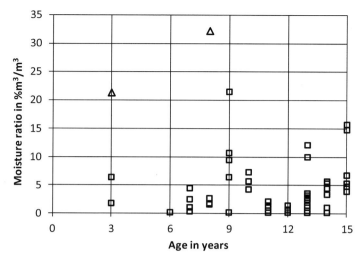

Figure 5.38. Protected membrane roof with unvented concrete tiles as ballast. Volumetric moisture ratio in the insulation boards depending on age. The triangles represent EPS, the squares XPS.

As long as the insulation boards are not ballasted with capillary tiles or wet soil, drying proceeds in summer, even in moderate climates. However, if ballasted, solar driven vapour flow from above the insulation and winter diffusion from the water puddles below alternate, causing interstitial condensation in the insulation year round, even with a bitumen lined underside. Of course, the amounts are still much lower in a vapour retarding than in a vapour permeable insulation material. But in the long run, also the retarding one will become wet, as the measured values of Figure 5.38 underline.

Apparently, although a vapour retarding insulation material increases moisture tolerance of a protected membrane roof, a long lasting guarantee can only be given if (1) the roof gets enough slope or a filter fabric below the ballast to avoid water puddling on the membrane and (2) the ballast stays neither moist by rain nor by under-cooling condensate.

5.4.2.3 Other performances

For most other performance requirements, a protected membrane roof scores excellently. Because the insulation is outside, thermal inertia of a heavy deck roof is high. The insulation also limits the temperature extremes in the membrane and protects it against UV. Where in a moderate climate membranes of compact low-sloped roofs experience –20 to 75 à 80 °C, the protected membrane roof insulation brings that interval down to 7 °C – 34 °C.

5.4.3 Design and execution

5.4.3.1 Roofing membrane

The excellent 'other performances' insinuate one could use membranes of lesser quality. This is untrue. To start with, their hygric and biological load scores higher than for compact roofs. Not only is the membrane permanently wet, its temperature is also quite stable, air is within reach and without filter fabric seeping rain carries along quite some organic dust. In other words,

5.4 Protected membrane roofs

these are ideal conditions for biological activity. Also finding leaks in case of rain penetration is an onerous job. Before any search can start, ballast, filter fabric if present and the insulation have to be removed! Both facts require identical membrane quality as for compact roofs: (1) double-layer, (2) top layer in polymer bitumen, (3) no organic inserts.

5.4.3.2 Details

A thermal bridge free design is somewhat more difficult for protected membrane roofs than it is for compact roofs. Take the edges. A thermal cut or wrapping up with insulation is needed. This is frequently forgotten. Indeed, after water proofing, the insulation comes, although wrapping up the edges should have been done before. That way, the activity 'insulating' splits into two rounds: edges and roof surface, enough to forget the edges.

5.4.3.3 Globally

Minimizing risks with protected membrane roofs requires:

1. A roof surface with enough slope, minimum 1.8%
2. Thermal bridge inhibiting details (edges, building-ups, etc.)
3. A membrane of excellent quality
4. A non-capillary, vapour retarding, water head and frost resistant isolation material (XPS)
5. The insulation boards well pushed together
6. A filter fabric upon the insulation as to assure maximum rain drainage there
7. A non-capillary ballast

Rule 7 conflicts with green roofs. If anyhow applied on a protected membrane roof, the solution requires an excellent, if possible vented drainage layer above the filter membrane. Also tiled ballasts and parkings challenge rule 7. Tiles must allow venting below (Figure 5.39), while parkings desks should have a vented drainage layer above the filter fabric.

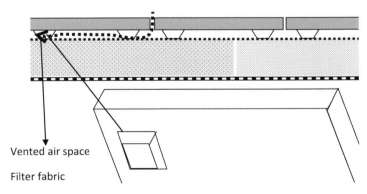

Figure 5.39. Protected membrane roof with vented tiles as ballast.

5.5 References and literature

[5.1] Dächer, Dachdeckungen (1965). DBZ, 12/1965 (in German).

[5.2] Hens, H. (1971). *Nota's bij de cursus burgerlijke bouwkunst* (in Dutch).

[5.3] Van der Kooy, J. (1971). *Moisture transport in cellular concrete roofs*. Delft, Uitgeverij Waltman, 105 pp.

[5.4] Berthier, J., Delcelier, P. (1972). *Etude experimentale des mouvements d'humidité dans une dalle de toiture*. Cahiers du CSTB, livraison 122, Cahier 1136, 12 pp. (in French).

[5.5] Isaksen, T., Helge, J. (1974). *Bitumen felt roofings on polystyrene insulated roofs*. Paper 2.5.1. in the Proceedings of the 2nd International CIB-RILEM conference on Moisture Problems in Buildings, Rotterdam.

[5.6] Gertis, K., Hauser, G. (1975). *Temperaturbeanspruchung von Stahlbetondächern*. IBP-Mitteilung 10 (in German).

[5.7] Hens, H. (1975). *Theoretische en experimentele studie van het hygrothermisch gedrag van bouw- en isolatiematerialen bij inwendige condensatie en droging met toepassing op de platte daken*. Doctoraal proefschrift, KU Leuven, 311 pp. (in Dutch).

[5.8] WTCB (1980). *Technische Voorlichting 134, Bepalen van de dakopbouw, uitgaande van hygrothermische gegevens*. Hellende daken, platte daken (in Dutch).

[5.9] Carpentier, G., De Kesel, J. P., Hens, H., Uyttenbroeck, J., Vaes, F. (1982). *Onderzoek van het hygrothermische gedrag van lichte daken met bitumineuze afdichting*. WTCB-tijdschrift, No. 1 (in Dutch).

[5.10] Künzel, H. (1987). *Keine Dampfsperre zwischen Dämmstoff und Gasbeton*. Das Dachdeckerhandwerk, 108, Heft 19, pp. 14–18 (in German).

[5.11] Hens, H. (1987). *Buitenwandoplossingen voor de residentiële bouw: platte daken*. Nationaal Programma RD Energie, 150 pp. (in Dutch).

[5.12] Laboratorium Bouwfysica (1989). *TV 134, Tekstvoorstel voor een nieuwe uitgave*, Rapport 89/23, 41 pp. (in Dutch).

[5.13] WTCB, Kursus-Conferentie No. 60, Platte Daken (1990). 63 pp. (in Dutch).

[5.14] König, N. (1990). *Wärmeschutz von Dachbegrünungssystemen – welche Schichten sind anrechenbar*. IBP-Mitteilung 197 (in German).

[5.15] WTCB (1992). *Technische Voorlichting 183, Het platte dak, opbouw, materialen, uitvoering, onderhoud*, 64 pp. (in Dutch).

[5.16] TI-KVIV, Studiedag Platdakrevolutie, Antwerpen, 18 november 1993 (in Dutch).

[5.17] WTCB (1994). *Technische Voorlichting 191, Het platte dak, aansluitingen en afwerking*, 91 pp. (in Dutch).

[5.18] Kyle, D., Desjarlais, A. (1994). *Assessment of Technologies for Constructing Self-Drying Low Slope Roofs*. ORNL/CON-380, 114 pp.

[5.19] Hens, H. (1995). *Thermal Performance of Protected Membrane Roof Systems. Part 1: A Simplified Heat and Moisture Analysis*. Journal of Thermal Insulation and Building Envelopes, Vol. 18, pp. 377–389.

[5.20] Künzel, H. M., Holz, D. (1995). *Assessing precipitation heat losses of inverted roofs*. IEA-Annex 24, Report T5-D-95/01.

[5.21] Homb, A., Store, M. (1995). *Moisture content in inverted roof constructions, results from a test house*. IEA-Annex 24, Report T5-N-95/01.

5.5 References and literature

[5.22]   Steiner, P. (1995). *First Results of a Study about Inverted Roofs with Improved Thermal Performance*. IEA-Annex 24, Report T5-CH-95/01.

[5.23]   Hens, H. (1996). *Thermal Performance of Protected Membrane Roof Systems. Part 2: Experimental Verification*. Journal of Thermal Insulation and Building Envelopes, Vol. 19, April, pp. 314–335.

[5.24]   Meert, E., Vitse, P. (1996). *Naar een duurzaam plat dak*. WTCB-tijdschrift No. 1 (in Dutch).

[5.25]   Vitse, P., Meert, E., Dubois, J. (1999). *Dakafdichtingen en dakisolatie: gebruik van bitumineuze koudlijmen en de ATG-aanpak*. WTCB-tijdschrift No. 3 (in Dutch).

[5.26]   WTCB (2000). *Technische Voorlichting 215, Het platte dak, opbouw, materialen, uitvoering, onderhoud*, 100 pp. (in Dutch).

[5.27]   Leimer, H., Rode, C., Künzel, H., Bednar, T. (2005). *Requirements for inverted roofs with a drainage layer*. Proceedings of the Vth Nordic Conference on Building Physics, Reykjavik (Ed.: G. Johannesson), pp. 570–577.

[5.28]   Uvslokk, S. (2005). *Calculation of moisture and heat transfer in compact roofs and comparison with experimental data*. Proceedings of the Vth Nordic Conference on Building Physics, Reykjavik (Ed.: G. Johannesson), pp. 633–640.

[5.29]   Uvslokk, S. (2005). *Calculation of moisture and heat transfer in compact roofs and comparison with experimental data*. Proceedings of the Vth Nordic Conference on Building Physics, Reykjavik (Ed.: G. Johannesson), pp. 633–640.

[5.30]   Bianchi, M., Desjarlais, A., Miller, W., Petrie, T. (2006). *Evaluating the energy performance of ballasted roof systems*. Research in Building Physics and Building Engineering (Eds.: P. Fazio, H. Ge, J. Rao, G. Desmarais), pp. 473–478.

[5.31]   Hens, H., Vaes, F., Janssens, A., Houvenaghel, G. (2007). *A Flight over a Roof Landscape: Impact of 40 Years of Roof Research on Roof Practices in Belgium*. Proceedings of Buildings X, Clearwater beach, December 2–7 (CD-Rom).

[5.32]   Rose, W. B. (2007). *White Roofs and Moisture in the US Desert Southwest*. Proceedings of Buildings X, Clearwater beach, December 2–7 (CD-Rom).

[5.33]   Bianchi, M., Desjarlais, A., Miller, W., Petrie, T. (2007). *Cool Roofs and Thermal Insulation: Energy Savings and Peak Demand Reduction*. Proceedings of Buildings X, Clearwater beach, December 2–7 (CD-Rom).

[5.34]   Holme, J., Noreng, K., Kvande, T. (2008). *Moisture and mold growth in compact roofs-Results from a three-stage field study*. Proceedings of NSB2008, Copenhagen, June 16–18, pp. 1221–1228.

[5.35]   Ray, S., Glicksman, L. (2010). *Potential Energy Savings of Various Roof Technologies*. Proceedings of Buildings XI, Clearwater beach, December 5–9 (CD-Rom).

[5.36]   Nusser, B., Bednar, T., Teibinger, M. (2011). *Proposal for a modified Glaser-Method for the risk assessment of flat timber roofs*. Proceedings of NSB2011, Tampere, May 29-June 2, pp. 181–188.

6 Pitched roofs

6.1 Classification

The classification of pitched roofs is based on form, supporting structure, and covering.

6.1.1 Type of form

We differentiate between simple and composite pitched roofs. The simple ones consist of one roof volume. The composite ones combine several intertwined roof volumes, a situation seen with complex floor plans.

6.1.1.1 Simple roofs

Classification is based on the longitudinal and transverse sections. The latter discriminates between saddle, French, mono-pitch roofs or double-slope pitch roofs. The first pertains to gable, false hip, hipped or tent roofs, see Figure 6.1. Detailing difficulty increases when moving from gable to tent. The ridge, the gutter, and the junction with the end walls are complex in a gable roof. False hip and hipped roofs add the intersection between pitches, called the roof strings. With tent roofs, the rooftop creates extra complexity.

Transverse → Longitudinal ↓	Saddle	French	Mono-pitch	Double sloped
Gable	X	X	X	X
False hip	X	X	X	X
Hipped	X	X	X	X
Tent	X	X	X	X

Figure 6.1. Pitched roof: classification array.

Performance Based Building Design 2. From Timber-framed Construction to Partition Walls.
First Edition. Hugo Hens.

6.1.1.2 Composite roofs

Composite roofs combine convex and concave intersections between simple roof volumes. While the convex ones give roof strings, the concave ones create valley gutters. These may collect such important amounts of rainwater during precipitation that they have to be rain tight (Figure 6.2).

Figure 6.2. Composite pitched roof, valley gutter.

6.1.2 Type of supporting structure

Purlins and rafters are most common as supporting structures. Rarer ones are catalogued as 'other'.

6.1.2.1 Purlins

The supporting structure exists of (Figure 6.3):

- Roof trusses, centre-to-centre distance 3 to 4 m. With small spans, triangular trusses or partitions walls bear the purlins. Larger spans demand trussed or composite girders. Girder slope establishes the roof slope.
- On the trusses purlins. These run normal to the slope, 1 to 2 m centre-to-centre. The ridge gets a sub-purlin. Cross sections: 130×63 to 225×75 mm.
- Ribs in case of small format coverings. These are mounted on the purlins parallel to the slope, 0.4 to 0.6 m centre-to-centre. The cross section varies between 65×65 and 95×65 mm. In many new roofs, planks with height equal to the insulation thickness are used.

6.1.2.2 Rafters

Here, slender rafters, centre-to-centre 0.6 to 0.8 m replace the supporting structure formed by trusses, purlins, and ribs (Figure 6.3). For small spans, triangular rafters suffice, while large spans demand trussed girders.

6.2 Roof covers in detail

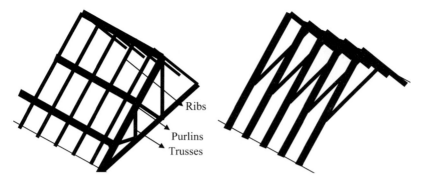

Figure 6.3. Supporting structure: purlins (to the left) and rafters (to the right).

6.1.2.3 Other

For very large spans steel trusses are used. As a rule, flat-trussed girders but sometimes space trusses apply. Fire safety may require the construction of a pitched roof in reinforced concrete. To increase temperature damping of the pitches, a good choice are steel, concrete or glued timber trusses finished with cellular concrete floor units.

6.1.3 Type of cover

Tiles, slates, corrugated sheets, and shingles find usage as covers. Interlocking, stemming run-off and pushed-up rain, makes tiled covers rain tight. Overlappings, whose height passes backpressure at design wind and capillary rise, assure rain-tightness of slated and corrugated sheet coverings. With shingles, the roof gets an OSB or plywood sheathing, which is finished with a bituminous membrane before nailing the shingles in overlapping rows.

6.2 Roof covers in detail

6.2.1 Tiles

We distinct between ceramic, concrete, and metal tiles.

6.2.1.1 Ceramic

The raw material for ceramic tiles is weathered clay. After mixing, the clay is pressed in the intended tile form, dried and then burned at high temperature. The tile name reflects the form: roman tile, pan tile, flat, etc. (Figure 6.4), which all can have single, double, or triple interlock. Each type differs in slope scheme (Figure 6.5). Cone tiles cover the strings and ridges.

Mounting a tiled cover goes down up. The roofer hooks each tile behind the laths in overlap with the one below and interlocking with the one aside. In case wind suction with a return period of 65 years exceeds the tile's weight per m², normal to the pitch, 1, 2 or 4 on 4 are nailed or clipped at the laths, see Table 6.1 and equation (6.1):

$$G_{\text{tile}} \left(= 9.81 \, m_{\text{tile}} \cos(h)\right) \leq p_{\text{w}} \left(= 0.27 \, C_{\text{p}} \left(1.3 \, q_0\right)\right) \rightarrow \text{nailing or clipping} \qquad (6.1)$$

Figure 6.4. Roman tile with single interlock to the left, pan tile with double interlock to the right).

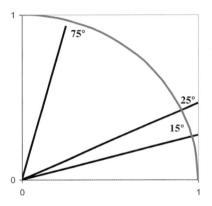

Slope below 15°, do not apply
Slope between 15 and 25°, combine with rain-tight underlay
Slope above 25°, clipping 1, 2 or 4 tiles on 4 where and when needed

Figure 6.5. Ceramic tiles, slope scheme.

In this inequality, m_{tile} is tile weight in kg/m², h roof slope, C_p local wind suction factor, q_0 dynamic wind pressure, 1.3 a multiplier considering the 65 years return period and 0.27 a factor accounting for pressure relief across a tiled cover. Wind suction is greatest along the roof edges and at the leeside of chimneys and other obstacles, see Table 6.2. The ratio between tile weight per m² and wind suction (G_{tile}/p_0) defines if 1, 2 or 4 tiles on 4 have to be clipped.

Table 6.1. Ceramic tiles, clipping.

Ratio Tile weight/wind suction	Clipping?
$0.8 \leq 1$	1 tile on 4
$0.6 \leq 0.8$	2 tiles on 4
< 0.6	4 tiles on 4

6.2 Roof covers in detail

Table 6.2. Wind suction with a return period of 65 years on the tiles ($p_w = 0.27\ C_p\ (1.3\ q_0)$).

Zone	Slope, °	Ridge height in m									
Coastal areas								7.0	9.0	11.5	14.5
Rural areas		5.0	6.0	7.5	9.5	12.0	14.0	18.0	22.0	27.0	32.0
Urbanized area		10.0	11.0	13.0	16.0	19.0	23.0	27.0	32.0	40.0	46.0	54.0
City centres		18.0	19.5	22.0	26.0	32.0	37.0	42.0	50.0	57.0	66.0	76.0
		Wind suction, Pa									
Corners (part of roof border)[1]	15 to 25	−555	−576	−614	−658	−702	−746	−790	−834	−878	−921	−965
	> 25: see edges											
Gutters (part of roof border)	15 to 25	−267	−276	−295	−316	−337	−358	−379	−400	−421	−442	−463
	> 25: see edges											
Edges (part of roof border)	15 to 25	−444	−461	−491	−527	−562	−597	−632	−667	−702	−737	−772
	25 to 50	−333	−345	−369	−395	−421	−448	−474	−500	−527	−553	−579
	> 50	−267	−276	−295	−316	−337	−358	−379	−400	−421	−442	−463
Roof centre	15 to 25	−222	−230	−246	−263	−281	−298	−316	−333	−351	−369	−386
	25 to 50	−267	−276	−295	−316	−337	−358	−379	−400	−421	−442	−463
	> 50	−222	−230	−246	−263	−281	−298	−316	−333	−351	−369	−386

[1] Roof border: its width a is found by first drawing the enveloping rectangle with sides b_1 and b_2 around the floor plan. If the ridge height is h and b_2 the smallest of the sides, width a is then given by:

b_2	a (m)
$\leq 3h$	Max $(1, 0.15\ b_2)$
$> 3h$	Max $(1, 0.45\ h, 0.04\ b_2)$

6.2.1.2 Concrete

Manufacturing concrete tiles consists of pressing fresh concrete into tile shapes and subjecting these to accelerated autoclave binding. Mostly used are profiled tiles, which fit in a 30 by 30 cm raster. Enough overlapping between the successive rows and interlocking with the tiles aside guarantees rain-tightness. Also here, the tiles are hooked behind the laths and additionally clipped depending on wind suction (Table 3.2, Figure 6.6). Cone tiles cover the strings and ridges.

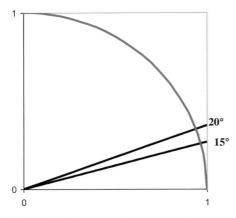

Figure 6.6. Concrete tiles, slope scheme.

Slope below 15°, do not apply
Slope between 15 and 20°, combine with a rain-tight underlay
Slope above 20°, clipping 1, 2 or 3 tiles on 4 where and when needed

6.2.1.3 Metallic

Metallic tiles are made of steel sheets, punched into three or four tile shapes aside each other. The roofer mounts the sheets bilaterally overlapping on laths and fixes them mechanically. The overlaps care for rain-tightness. Cone tiled steel sheets cover the strings and ridges.

6.2.2 Slates

6.2.2.1 Quarry

Manufacturing quarry slates includes cleaving slate rocks and sawing the sheets obtained into shape. Popular forms are flat and diamond. Slates are mounted on a plank lining or on laths with each following row staggered half a slate and overlapping the one below so that three slates lie on top of each other everywhere, see Table 6.3 and Figure 6.7. As the table indicates, slates are nailed or hooked. The low (L), moderate (M) and high wind (H) driven rain indexes mentioned are found by multiplying annual precipitation in m per year (G_{Rh}) by wind pressure $p_{w,basis}$ with a return period of 10 years. Preformed zinc sheets cover the strings and ridges.

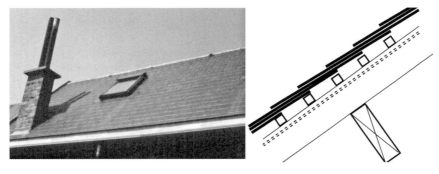

Figure 6.7. Slates, nailed in overlap on laths.

6.2 Roof covers in detail

Table 6.3. Overlap between each first and third row of slates.

Pitch, horizontal projection		Overlap l in mm									
		Nailing the slates					Hooking the slates				
		Wind-driven rain index					Wind-driven rain index				
< 5.5 m		L	M	H			L	M	H		
5.5 to 11 m			L	M	H			L	M	H	
> 11 m				L	M	H			L	M	H
Slope (cm/m)	30	133	139	157	174	192					
	40	102	116	129	143	156	126	140			
	50	91	102	113	124	136	106	117	128	139	150
	70	79	87	96	105	113	83	92	101	110	118
	90	72	80	87	95	102	72	80	87	95	102
	110	69	76	82	89	96	64	73	79	86	93
	130	67	73	79	85	92	60	68	74	91	87
	150	65	71	77	83	89	57	65	71	77	83
	200	63	69	74	80	85	54	61	66	72	78
	300	61	67	72	77	82	51	58	63	68	74
	> 1000	60	65	70	75	80	50	55	60	65	70

Wind -driven rain index:

$G_{Rh} \times p_{w,basis}$	Classification
≤ 600 Pa·m	Low (L)
600–1200 Pa·m	Moderate (M)
> 1200 Pa·m	High (H)

6.2.2.2 Fibre cement

Actual fibre-cement contains cellulose fibres. This increases hygric movement compared to the former asbestos-based fibre cement slates. Shape choice, mounting and overlapping does not differ from quarry slates, see Table 6.3. Preformed zinc sheets cover the ridges.

6.2.2.3 Timber slates and plain tiles

Cleaving tree-trunks lengthwise into planks and sawing up these in slates gives timber slates. Plain tiles instead consist of fired clay with all irregularities this gives. Plain tiles exist in different shapes. Timber slates and plain tiles are usually smaller than quarry and fibre-cement slates. They are mounted the same way. Preformed zinc sheets cover the strings and ridges.

6.2.3 Corrugated sheets

One has fibre-cement or synthetic corrugated sheets. Their shape gives good bending strength, allowing spans that make laths redundant. Correct overlapping between successive rows cares for rain-tightness. Sheets aside each other overlap corrugation on corrugation. Each sheet is fastened by screwing on top. For the slope scheme, see Figure 6.8. Preformed sheets cover the string and ridges.

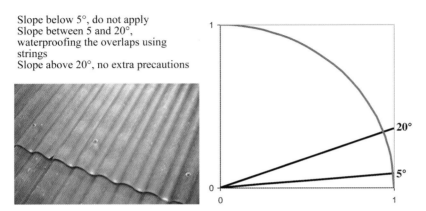

Figure 6.8. Corrugated sheets, screwing and slope scheme.

6.2.4 Shingles

Whereas shingles are not popular in North-Western Europe, in North America they are the most common pitched roof cover. Manufacturing includes cutting coated bitumen with glass fibre insert in sheets with dimensions ±40 × 100 cm and making three equidistant halfway incisions in the sheets to form four slate-like shingle shapes. The sheets are laid side by side and nailed across a coated bitumen underlay into an OSB or plywood lining. Each following row is staggered half a shingle compared to and overlaps the one below so that three shingles lie on top of each other everywhere (Figure 6.9).

Figure 6.9. Shingles, overlaps and nailing.

6.3 Basic assemblies

The simplest assembly consists of a supporting structure (purlins with ribs or rafters) with tile laths, slate laths or timber lining on top, finished with the roof cover chosen (Figure 6.10(a)). Lath cross section depends on the span:

Ribs or rafters, distance centre-to-centre mm	Laths, cross section $b \times h$ in mm²
400	26 × 32
500	26 × 40
600	36 × 40

Such simple assembly however, suffers from air leakage and easy dust penetration, powder snow seeping and windstorm damage sensitivity. Adding an underlay limits these disadvantages (Figure 6.10(b)). From the bottom up we then have: supporting structure; either underlay, battens and laths, or a timber, OSB or plywood lining; roof cover. The battens are nailed on the ribs or rafters across the underlay. They run parallel to the slope, allowing the underlay to act as secondary rain screen. Moreover, they facilitate wind washing and venting below the cover. Battens are half to one lath high as higher may weaken wind stability of the roof cover.

Underlays used are: coated bitumen on a timber lining (extremely vapour tight), aluminium button foils (vapour tight), micro-perforated glass fabric reinforced synthetic foils (vapour retarding), thin fibre cellulose cement sheets (vapour permeable, capillary), fibreboard (vapour permeable) or spun bonded synthetic foils (highly vapour permeable).

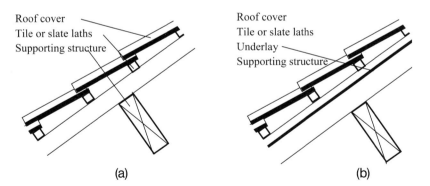

Figure 6.10. Simplest assembly (a), underlay added (b).

The assembly with underlay refers to uninhabited attics, where the thermal insulation sits at ceiling level, see Figure 6.11. If instead the attic belongs to the inhabited space, the insulation moves to the pitches, a solution called 'cathedralized ceiling'. In such a case, the insulation is located on, in between or under the ribs or rafters, see Figure 6.12. If on, the insulating roof decking or insulation boards function as underlay. If in-between, mineral wool or cellulose fibre fills the bays formed by ribs or rafters. If under, then stiff insulation boards close the space taken by the ribs or rafters. In all three cases, the pitches get a lathed ceiling, gypsum board lining, or sprayed gypsum plaster on expanded metal mesh as inside finish.

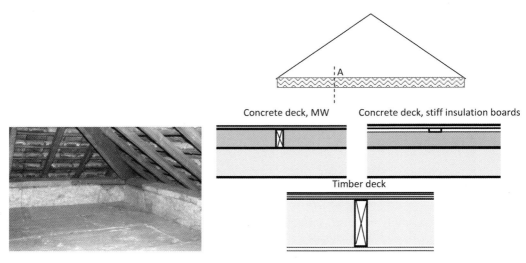

Figure 6.11. Uninhabited attic, insulation at ceiling level.

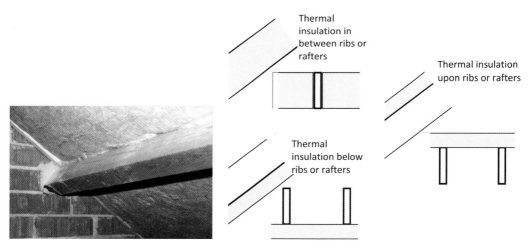

Figure 6.12. Cathedralized ceiling, insulation on, in between, or under the ribs or rafters.

6.4 Performance evaluation

6.4.1 Structural integrity

Trusses and rafters give lateral thrust, which the ceiling deck (Figure 6.13) or the supporting structure must neutralize. A solution exists of adding tirants to the trusses or rafters. The principals of large span trusses received extra support beams under the purlins in the past, resulting in complex geometries with an extended vocabulary to indicate the parts.

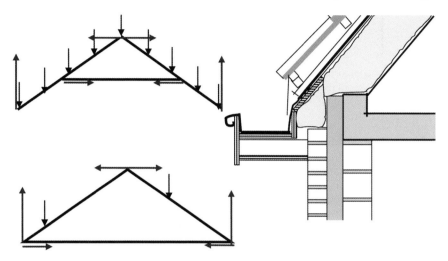

Figure 6.13. Lateral thrust neutralized by the ceiling deck detailed.

Where purlins and ribs cross, the vertical load by own weight and dead load decomposes in a component normal and parallel to the pitch. The normal one bends the purlins, the parallel one induces traction in the ribs if coupled above the subpurlin at the ridge. The ribs compress if supported by the ceiling deck. When coupled above and supported below then both traction and compression intervene. Traction results in extra loading of the subpurlin. Its cross section must account for that. Compression in turn gives lateral thrust on the façade. When the design does not consider these forces, the building may burst with crack formation in the cross and gable end walls. The more the façade walls push away, the more the subpurlin gets loaded. It's bending increases, as does pushing. The problem disappears when with a stiff underlay the pitch works as a diaphragm. This conducts the parallel component to the trusses and gable end walls (Figure 6.14).

Also, wind load demand the necessary precautions. While a truss is stable in its plane, it is not normal to that plane. A system of trusses, but also of trusses, purlins and ribs behaves like a house of cards. Wind stability presumes the roof structure retains its form. Composite roofs where strings and valley gutters function as diagonals. In all other cases, diagonals must at least stiffen one bay or a stiff underlay must allow diaphragm action (see above).

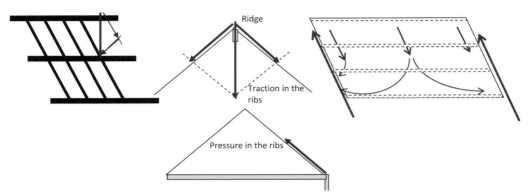

Figure 6.14. On the left decomposing the vertical load by own and dead weight in a component normal and parallel to the pitch, in the middle forces in ribs and subpurlin at the ridge, on the right diaphragm action thanks to a stiff underlay.

6.4.2 Building physics: heat, air, moisture

Pitched roofs are a story unto themselves. As long as thermal insulation was not an issue, they were hardly damage sensitive. Insulation changed this. Pitched roofs started dripping and showed timber rot. Frost damaged the tiles, etc. Not only was insulation to blame, the use of certain underlays and the transformation of uninhabited attics into living spaces also contributed. For a long time, builders sought a solution in combining two classic measures: a vapour retarding layer at the winter warm side of the thermal insulation and outside air ventilation between insulation and underlay. A vapour retarder was intended to prevent water vapour from diffusing into the assembly. The vapour that nevertheless entered the roof had to be evacuated by ventilation before condensing at the underlay's backside. None of the two measures was successful. Today, we know the culprits: uncontrolled air transfer in and across the pitched roof assembly and under-cooling of the cover by long wave radiation to the clear sky.

6.4.2.1 Air tightness

We make a difference between unused attics with the insulation at ceiling level and attics used as living spaces.

Insulation at ceiling level

When a ceiling without walked on finish above the insulation lacks air tightness, air from the living zone will infiltrate into the attic. Even if problem free at ceiling level, living space ventilation loses controllability because of this, while the probability of high attic relative humidity and condensation at the underlay or cover increases. With a walked on finish, compromised air tightness could also intensify air washing of the insulation.

As long as reinforced concrete decks, prefabricated structural floor units with concrete topping and timber floors lined at the underside with gypsum board or sprayed gypsum plaster on expanded metal mesh lack local leaks formed by the access panel perimeter, passages for ventilation pipes, electrical lines, and others, they are sufficiently air-tight. With such leaks, these decks need an air-retarding layer under the insulation, for example a 0.2 mm thick PE-foil with the overlaps taped (Figure 6.15).

6.4 Performance evaluation

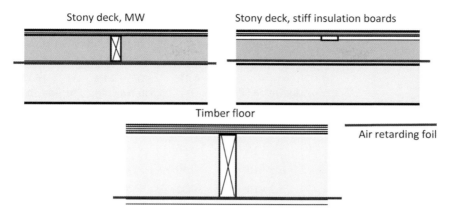

Figure 6.15. Uninhabited attic, insulation at ceiling level, inclusion of an airtight foil when necessary.

On a gypsum board lining, the foil should also cover the junctions with partition and outer walls as shrinkage often causes cracking there. Timber decks without or with a fractionated ceiling cannot without air retarding corrective measures.

Cathedral ceilings

Cathedral ceilings are critical as an assembly. Neither the roof cover nor the underlay and thermal insulation guarantee air-tightness. If the internal lining also does not, in- and exfiltration will occur. Lack of wind-tightness could in turn activate wind washing, while bad workmanship in mounting the insulation layer may induce air looping.

In- and exfiltration

Due to the air permeance of the layers composing the pitches, see Table 6.4, in- and exfiltration seem unavoidable. Only gypsum board is more or less airtight. Table 6.5 gives laboratory data for roofs without underlay. Carelessly mounting the glass fibre blankets and disregarding air tightness of the inside lining increases the air permeance coefficient with a factor 10, from $\approx 10^{-5}$ to 10^{-4} kg/(m² · s · Pa^{b-1}). We measured the same when testing on site, see Table 6.6. Air tightening a cathedral ceiling thus looks not as simple as pretended. Doing better than 10^{-5} kg/(m² · s · Pa^{b-1}) is hardly practical, although the performance criteria listed in literature are even more strict, see Table 6.7. This has consequences for design and execution.

In windy weather, an air permeable cathedral ceiling suffers from infiltration at the wind side and exfiltration at the leeward side. Winter thermal stack in turn activates infiltration at the bottom and exfiltration at the top of the pitches. When both act in common, infiltration develops at the wind side and at the leeward side's bottom while exfiltration concentrates at the leeward side's top. For pitches with little slope, wind and thermal stack turn the whole roof into an exfiltration plane. With in- and exfiltration, thermal transmittance loses its value as measure for 'insulation quality', whereas in winter, exfiltration negatively affects moisture tolerance. Infiltration instead may generate draught complaints, degrades transient response and lowers noise reduction.

Table 6.4. Air permeance: roof cover, underlay, insulation layer, inside lining.

Layer	Air permeance $K_a = a\,\Delta P_a^{b-1}$		In- or exfiltration at an air pressure difference of	
	a kg/(m²·s·Paᵇ)	$b-1$ —	$\Delta P_a = 2$ Pa m³/(m²·h)	$\Delta P_a = 10$ Pa m³/(m²·h)
Roof covers				
Roman tiles, single interlock	$1.6 \cdot 10^{-2}$	−0.50	68.0	152.0
Roman tiles, double interlock (di)	$1.3 \cdot 10^{-2}$	−0.50	55.0	114.0
Roman tiles, di, 1 venting tile per m²	$1.5 \cdot 10^{-2}$	−0.50	64.0	142.0
Pan tiles	$1.2 \cdot 10^{-2}$	−0.32	58.0	172.0
Concrete tiles	$7.8 \cdot 10^{-3}$	−0.46	49.0	81.0
Fibre-cement slates	$1.7 \cdot 10^{-3}$	−0.21	8.8	31.0
Quarry slates	$5.4 \cdot 10^{-3}$	−0.34	26.0	74.0
Metallic tiles	$2.1 \cdot 10^{-3}$	−0.43	9.4	23.4
Corrugated fibre cement sheets	$9.1 \cdot 10^{-4}$	−0.37	4.2	11.6
Underlay				
FCC[1]-sheets, overlap taped	$4.2 \cdot 10^{-4}$	−0.34	2.0	5.8
Ditto, overlap open (3.6 mm)	$3.2 \cdot 10^{-3}$	−0.40	15.0	38.0
Perforated glass fabric reinforced synthetic foil	$5.0 \cdot 10^{-3}$	−0.80	17.0	24.0
Thermal insulation layers				
Mineral wool, $\rho = 30$ kg/m³	$0.0200\,\rho^{-1.5}/d$	0	3.7	18.3
Glass fibre, $\rho = 15$ kg/m³	$0.0043\,\rho^{-1.3}/d$	0	3.8	19.1
XPS, $\rho = 32$ kg/m³, groove and tongue	$4.0 \cdot 10^{-4}$	−0.41	1.8	4.7
Ditto, no groove and tongue, closed	$5.4 \cdot 10^{-4}$	−0.46	2.4	5.6
Ditto, groove and tongue, 2 mm gap	$2.0 \cdot 10^{-3}$	−0.45	8.8	21.3
Internal linings				
Lathed ceiling, groove and tongue	$4.1 \cdot 10^{-4}$	−0.32	2.0	6.0
Ditto, leak ⌀ 20 mm for electric wiring	$7.6 \cdot 10^{-4}$	−0.37	3.5	9.7
Gypsum board, joints plastered	$3.1 \cdot 10^{-5}$	−0.19	0.16	0.6
Ditto, open joints	$3.3 \cdot 10^{-4}$	−0.39	1.5	4.0
Ditto, joints plastered, leak ⌀ 20 mm	$3.8 \cdot 10^{-4}$	−0.39	1.7	4.6
Alum-gypsum board, joints plastered	$1.3 \cdot 10^{-5}$	0	0.06	0.4
Ditto, joints plastered, leak ⌀ 20 mm	$4.7 \cdot 10^{-4}$	−0.47	2.0	4.8
Ditto, open joints	$5.6 \cdot 10^{-4}$	−0.41	2.5	6.5
Ditto, open joints, 3 mm wide	$3.0 \cdot 10^{-3}$	−0.43	13.0	33.0

[1] FCC = fibre-cement-cellulose

6.4 Performance evaluation

Table 6.5. Air permeance of cathedralized ceiling assemblies, laboratory measurements.

Assembly	Air permeance		In- or exfiltration at an air pressure difference of	
	a $kg/(m^2 \cdot s \cdot Pa^b)$	$b-1$ –	$\Delta P_a = 2$ Pa $m^3/(m^2 \cdot h)$	$\Delta P_a = 10$ Pa $m^3/(m^2 \cdot h)$
Assembly 1 (no underlay)				
1. Lathed ceiling	$3.0 \cdot 10^{-3}$	−0.37	14.0	38.0
2. 1 + mineral wool, $d = 10$ cm	$3.0 \cdot 10^{-3}$	−0.38	13.8	37.5
3. 2 + Roman tiles	$3.0 \cdot 10^{-3}$	−0.38	13.8	37.5
Assembly 2 (no underlay)				
1. Glass fibre blankets, carelessly mounted	$2.0 \cdot 10^{-3}$	−0.23	10.2	35.3
2. 1 + Roman tiles	$1.8 \cdot 10^{-3}$	−0.23	9.2	31.8
3. 2 + lathed ceiling	$2.7 \cdot 10^{-4}$	−0.46	1.2	2.8
4. 3 + perforation for electric wiring	$5.4 \cdot 10^{-4}$	−0.47	2.3	5.5
Assembly 3 (no underlay)				
1. Glass fibre blankets, correctly mounted	$1.4 \cdot 10^{-4}$	−0.39	0.6	1.7
2. 1 + lathed ceiling	$7.5 \cdot 10^{-5}$	−0.35	0.4	1.0
3. 2 + perforation for electric wiring	$1.0 \cdot 10^{-4}$	−0.40	0.5	1.2
Assembly 4 (no underlay)				
1. Alum glass fibre blankets, correctly mounted	$1.2 \cdot 10^{-5}$	−0.07	0.07	0.3
2. 1 + lathed ceiling	$7.1 \cdot 10^{-5}$	−0.39	0.3	1.0
3. 2 + perforation for electric wiring	$8.3 \cdot 10^{-5}$	−0.38	0.4	1.0
4. 3 + fissured alum vapour retarder	$1.7 \cdot 10^{-4}$	−0.38	0.8	2.1
Assembly 5 (lathed ceiling, alum glass fibre blankets, no underlay, Roman tiles, double interlock)				
1. Blanket flanges overlapping, taped	$1.6 \cdot 10^{-5}$	0	0.1	0.5
2. 1 + stapled tape	$1.6 \cdot 10^{-5}$	0	0.1	0.5
3. 1 + nail hole in the vapour retarder	$1.6 \cdot 10^{-5}$	0	0.1	0.5
4. 1 + fissured vapour retarder	$1.3 \cdot 10^{-4}$	−0.35	0.6	1.7
5. Flanges not overlapping	$1.6 \cdot 10^{-4}$	−0.30	0.8	2.4

Table 6.5. (continued)

Assembly	Air permeance		In- or exfiltration at an air pressure difference of	
	a kg/(m²·s·Pa^b)	$b-1$ —	$\Delta P_a = 2$ Pa m³/(m²·h)	$\Delta P_a = 10$ Pa m³/(m²·h)
Assembly 6 (lathed ceiling with leak $\varnothing$ 20 mm per m², alum glass fibre blankets, no underlay, Roman tiles, double interlock)				
1. Blanket flanges overlapping, taped	$1.5 \cdot 10^{-5}$	0	0.09	0.45
2. 1 + nail hole in the vapour retarder	$7.2 \cdot 10^{-5}$	−0.30	0.4	1.1
3. 1 + fissured vapour retarder	$2.3 \cdot 10^{-4}$	−0.42	1.0	2.6
4. Flanges not overlapping	$3.9 \cdot 10^{-4}$	−0.38	1.8	4.9
Assembly 7 (lathed ceiling, alum glass fibre blankets, no underlay, slates)				
1. Blanket flanges overlapping, taped	$1.1 \cdot 10^{-5}$	0	0.07	0.3
Assembly 8 (lathed ceiling, alum glass fibre blankets, no underlay, slates)				
1. Blanket flanges overlapping, taped	$1.1 \cdot 10^{-5}$	0	0.07	0.3
2. Flanges not overlapping	$3.9 \cdot 10^{-4}$	−0.40	1.8	4.9
Assembly 9 (gypsum board, alum glass fibre blankets, no underlay, tiles)				
1. Blanket flanges overlapping, taped	$1.2 \cdot 10^{-5}$	0	0.07	0.3
2. 1 + fissured vapour retarder	$1.1 \cdot 10^{-5}$	0	0.07	0.3
3. Flanges not overlapping	$1.0 \cdot 10^{-5}$	0	0.06	0.3
Assembly 10 (gypsum board, 1 leak $\varnothing$ 20 mm per m², alum glass fibre blankets, no underlay, tiles)				
1. Blanket flanges overlapping, taped	$1.2 \cdot 10^{-5}$	0	0.07	0.3
2. 1 + fissured vapour retarder	$1.8 \cdot 10^{-4}$	−0.35	0.8	2.4
3. Flanges not overlapping	$4.4 \cdot 10^{-4}$	−0.45	1.9	4.7
Assembly 11 (gypsum board with open joints, alum glass fibre blankets, no underlay, tiles)				
1. Blanket flanges overlapping, taped	$1.1 \cdot 10^{-5}$	0	0.07	0.3
2. Flanges not overlapping	$4.9 \cdot 10^{-4}$	−0.38	2.3	6.1
Assembly 12 (gypsum board with open joints, alum glass fibre blankets, no underlay, tiles)				
1. Blanket flanges overlapping, taped	$1.2 \cdot 10^{-5}$	0	0.07	0.3
2. 1 + fissured vapour retarder	$1.3 \cdot 10^{-4}$	−0.30	0.6	2.0
3. Flanges not overlapping	$8.8 \cdot 10^{-4}$	−0.43	3.9	9.8

'Alum' refers to glass fibre blankets with aluminium back-lining acting as vapour retarder.

6.4 Performance evaluation

Table 6.6. Air permeance of cathedralized ceiling assemblies, on site testing.

Assembly (down-up)	Air permeance		In- or exfiltration at an air pressure difference of	
	a $kg/(m^2 \cdot s \cdot Pa^b)$	$b-1$	$\Delta P_a = 2$ Pa $m^3/(m^2 \cdot h)$	$\Delta P_a = 10$ Pa $m^3/(m^2 \cdot h)$
Assembly 1				
Gypsum board PE-vapour retarder, perforated Glass fibre boards, $d = 15$ cm Vented air layer Plywood PVC-membrane (air-tight, roof edges are not)	3.2 à 3.6 · 10^{-4}	–1/3	1.6	4.5
Assembly 2				
Wood wool-cement boards PE-vapour retarder, leaking edges Glass fibre blankets, $d = 7.5$ cm Air layer Corrugated fibre-cement sheets	0.6 à 1.6 · 10^{-4}	–1/3	0.5	1.5
Assembly 3				
OSB-boards PE-vapour retarder, leaking edges Glass fibre blankets, $d = 10$ cm Synthetic underlay sheets Air space Enamelled corrugated steel sheets	3 à 3.8 · 10^{-4}	–1/3	1.6	4.5

Table 6.7. Air tightness requirements, Canada.

	Air flow at 75 Pa $kg/(m^2 \cdot h)$	Air permeance $kg/(m^2 \cdot s \cdot Pa)$	Diffusion thickness wind barrier m
Air retarder, material	< 0.084	< 0.36 · 10^{-6}	
Air retarder as mounted	< 1.040	< 3.24 · 10^{-6}	< 0.25
	< 0.216	< 0.84 · 10^{-6}	> 3.25
Wind barrier as mounted	< 4.500	< 16.8 · 10^{-6}	

Longitudinal flow

The word 'longitudinal' combines four flow patterns (Figure 6.16):

1. Outside airflow above the insulation, named 'venting' or, if purpose designed, 'ventilation'
2. Outside airflow in and under the insulation, named 'wind washing'
3. Inside airflow under the insulation
4. Inside airflow in and above the insulation, named 'inside air washing'

114　　　　　　　　　　　　　　　　　　　　　　　　　　　　　6 Pitched roofs

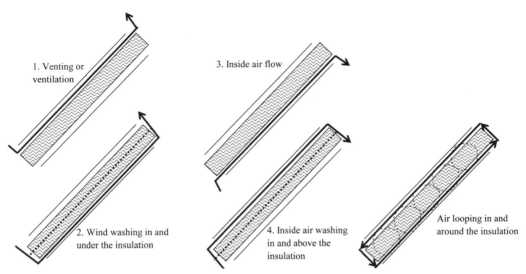

Figure 6.16. Cathedral ceilings, longitudinal airflow patterns, air looping.

Longitudinal flows develop with leaks at different heights in the layers in- and outside the insulation. This not only concerns cracks between gypsum boards, non-taped overlaps in the underlay and local perforations of the inside lining but also purpose designed ventilation in- and outlets. When the thermal insulation layer is leaky and an air layer wraps it, these four longitudinal flow patterns combine with air in- and exfiltration and air looping into one overall air movement picture.

Airflow patterns 1 and 3 are quite innocent. Pattern 1, venting, even figures as one of the classic measures mentioned to exclude interstitial condensation and facilitate drying. Patterns 2 and 4 on the contrary act negatively in terms of thermal performance. At the same time, winter wind washing lowers the inside surface temperatures and degrades the transient response in summer. Inside air washing for its part increases interstitial condensation risk in the roof.

Air looping

To see air looping develop in cathedralized ceilings, the following conditions have to be fulfilled (Figure 6.16): air layer at both sides of the thermal insulation, leaks at different height across this insulation or, an air permeable thick enough insulation layer. Air looping heightens the effective thermal transmittance and changes interstitial condensation patterns. Whereas diffusion and equally distributed exfiltration give an evenly spread deposit, with air looping the upper part of the pitches sees the most with a slow spread down when looping intensifies

6.4.2.2　Thermal transmittance

Calculating thermal transmittances (that adjective 'whole' is no longer repeated) is of use for airtight assemblies. In all other cases, 'effective' values, which currently are unknown, prevail. The word 'effective' relates to the mean heat flow crossing an assembly per unit of surface and time at a temperature difference of 1 degree centigrade between the environments at both sides, what the driving forces may be. Although each cathedralized ceiling should be acceptably airtight, this is a pious wish in almost all cases, which is why, after a short discussion about the insulation at ceiling level. Both airtight and air permeable are considered.

6.4 Performance evaluation

Insulation at ceiling level

In case the deck is acceptably airtight, thermal transmittance writes as $U_r = aU$, with a a multiplier accounting for the attic temperature being different from outdoors. For well-insulated decks, a equals 1 and $U_r = U$. With the insulation boards on the deck, a thermal transmittance of 0.1 to 0.4 W/(m² · K) results in the insulation thicknesses of Table 6.8. Table 6.9 gives the thickness when the insulation sits between the 20 cm high joists of a timber ceiling and a value 0.4 W/(m² · K) is the objective, while Table 6.10 lists the values reached for the bays fully filled with mineral wool or cellulose.

Table 6.8. Insulation on the ceiling deck, insulation thicknesses.

Ceiling deck	U	Insulation thickness in cm			
	W/(m² · K)	MW	EPS	XPS	PUR
Concrete, d = 14 cm, no screed, plastered inside	0.4	8	8	7	6
	0.2	17	17	16	13
	0.1	35	35	35	35
Hollow concrete floor units, d = 14 cm, no screed, plastered inside	0.4	8	8	7	6
	0.2	17	17	16	13
	0.1	35	35	35	35
20 cm high, 4 cm wide timber joists centre to centre 40 cm, OSB, gypsum board lining	0.4	7	7	7	5
	0.2	16	16	15	12
	0.1	35	35	32	26

Table 6.9. Timber ceiling, insulation between the joists. Thickness for $U = 0.4$ W/(m² · K).

Joists			Thickness in cm		
c to c[1] cm	Height cm	Width cm	MW	EPS	PUR
40	20	8	8	8	6
40	20	4	8	8	6
60	20	8	8	8	6
60	20	4	7	7	6

[1] c to c = centre to centre

Table 6.10. Timber ceiling, insulation between joists, complete fill.

Joists			U-value W/(m² · K)		ψ W/(m · K)
c to c[1] cm	Height cm	Width cm	MW	Cellul.	
40	20	8	0.25	0.25	0.028
40	20	4	0.22	0.22	0.014
60	20	8	0.23	0.23	0.028
60	20	4	0.21	0.21	0.014

A value 0.4 W/(m$^2 \cdot$ K) is easily met. 0.2 W/(m$^2 \cdot$ K) is somewhat more demanding with thicknesses ranging from 12 to 18 cm. The packages required to meet 0.1 W/(m$^2 \cdot$ K), however, generate secondary costs in terms of extra construction height needed. Due to the linear thermal transmittances (ψ) of the joists, filling the bays between 20 cm high joists with mineral wool or cellulose just misses 0.2 W/(m$^2 \cdot$ K), see the last column of Table 6.10.

In case of an air permeable ceiling deck, air leakage mostly happens through local leaks. Consequently, attic temperature increases a little whereas the conduction losses across the deck decrease somewhat, i.e., $a < 1$. Of course air leakage activates infiltration related ventilation and related heat losses. When instead the ceiling deck shows equally distributed air leakage over its complete surface, then the conduction related thermal transmittance changes from interface to interface. Logging at the inside surface gives too low, logging at the attic side too high thermal transmittances. An on-site test, with heat flow meters and thermocouples at the inside surface for example gave the following result:

Thermal transmittance, W/(m$^2 \cdot$ K)	
Calculated	**Measured**
0.55	0.38

Cathedralized ceilings

For an imposed thermal transmittance requirement, airtight assemblies demand insulation thicknesses hardly different from those at ceiling level. The insulation between ribs or rafters results in a parallel circuit of area weighted thermal resistances, one for the insulated bays and another for the ribs or rafters, in series with the remaining layers composing the assembly. Using the linear thermal transmittances given in Table 6.10 figures is an alternative. Glass fibre, mineral wool, and cellulose are the most appropriate insulation materials. For solutions with the insulation on or under the ribs or rafters, several materials fit: stiff glass fibre or mineral wool boards, EPS, XPS, PUR, CG.

As for the effect of in- and exfiltration, indoor air washing, wind washing, and air looping on the effective thermal transmittance, cathedral ceilings composed of air permeable layers behave the way cavity walls do. Underlay and roof cover replace the veneer, while the inside lining with leaky or tight air and vapour retarder replaces the inside leaf. Leakage likelihood, however, is higher. Whereas infiltration increases and exfiltration decreases conduction losses at the inside surface, wind washing, indoor air washing, and air looping uplift the mean 'effective' thermal transmittance, though with varying heat flow rate along the pitches. Although these phenomena are quantifiable using appropriate calculation models, experimental evaluation remains more convincing.

In- and exfiltration experimentally

Roof assembly 1

A first roof assembly tested in a hot box/cold box separated a 20 °C inside from a 1.5 °C outside environment. From inside to outside, the assembly had the following features (see Figure 6.17):

- Lathed groove and tongue ceiling
- Air cavity, $d = 20$ mm
- Glass fibre blankets, 54 mm thick, with as vapour retarding layer aluminium coated craft paper. The flanges did not overlap at the ribs
- Air cavity, $d = 90$ mm
- Ceramic Roman tiles

6.4 Performance evaluation

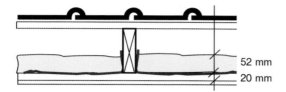

Figure 6.17. Cathedralized ceilings, assembly 1.

The roof had a heat flow meter at the inside surface and thermocouples at the in- and outside surface and at all interfaces in between. In a first step, testing happened with intact ceiling and an air overpressure in the hot box of 1.5 Pa (stage 1). In a second stage, a hole with a 20 mm diameter was drilled across the lathed ceiling, which reduced overpressure in the hot box to 1 Pa. In a third stage, overpressure went to 7 Pa. Thermal transmittances noted at the inside surface were:

Thermal transmittance at the inside surface, W/(m² · K)		
stage 1	stage 2	stage 3
0.36	0.36	0.24

Calculated thermal transmittance, using the conduction equation, gave 0.54 W/(m² · K). Or, the roof in stage 3 apparently contained 3 times more insulation then added. Temperature profiles in turn blew up with higher air outflow, resulting in less conduction at the inside surface, explaining the low thermal transmittance measured (Figure 6.18). If the heat flow had been logged at the outside surface, thermal transmittance should have increased instead. So, exfiltration clearly outperforms thermal transmittance as a single-valued cathedral ceiling property.

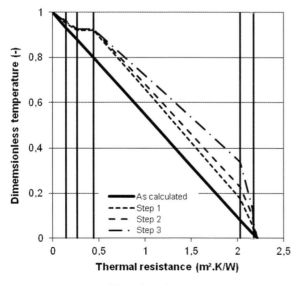

Figure 6.18. Assembly 1, interface temperatures as measured.

In- and ex-filtration, wind washing and air looping experimentally

Roof assembly 2, 3, and 4

A second hot box/cold box round included three roof assemblies. Assembly 2 was in to out composed of (Figure 6.19):

- Gypsum board lining (quite vapour permeable but strongly airtight)
- Service cavity, $d = 22$ mm
- Glass fibre, $d = 120$ mm, density 17 kg/m³ (bays between ribs fully filled)
- Fibre cement cellulose underlay (FCC), $d = 3.2$ mm, strip width 120 cm. FCC combines good vapour permeability ($\mu d = 0.14$–0.25 m) with moderate capillarity ($A = 0.011$ kg/(m² · s$^{0.5}$)). The overlaps between the strips make the underlay quite air permeable
- Laths and battens
- Glazed ceramic tiles with double interlock

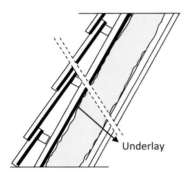

Figure 6.19. Roof assembly 2.

Assembly 3 mirrored 2, though without underlay, while the only difference with 2 in assembly 4 was the perforated, glass fabric reinforced synthetic underlay instead of FCC, in turn composed of 120 cm wide strips, be it with high vapour resistance now ($\mu d = 9$ m with standard deviation 4.8 m) but moderate air tightness thanks to the overlaps.

The three roofs faced ±22 °C in the hotbox and 2.8 °C in the cold box. In a first stage, the air pressure difference between hot and cold box was zero. In a second stage, overpressure in the hot box increased to 18 Pa. In stage three, overpressure went back to 2 Pa, but the gypsum board internal lining got two 4 mm wide crosscuts normal to the slope. Figure 6.20 shows the temperatures as measured in the three roof assemblies during each stage, field 1 representing assembly 2, field 2 assembly 3 and field 3 assembly 4. The picture suggests air looping around the insulation, perhaps with some exfiltration. In step 3, air looping overwhelms exfiltration with much higher average underlay temperatures as a direct consequence.

6.4 Performance evaluation

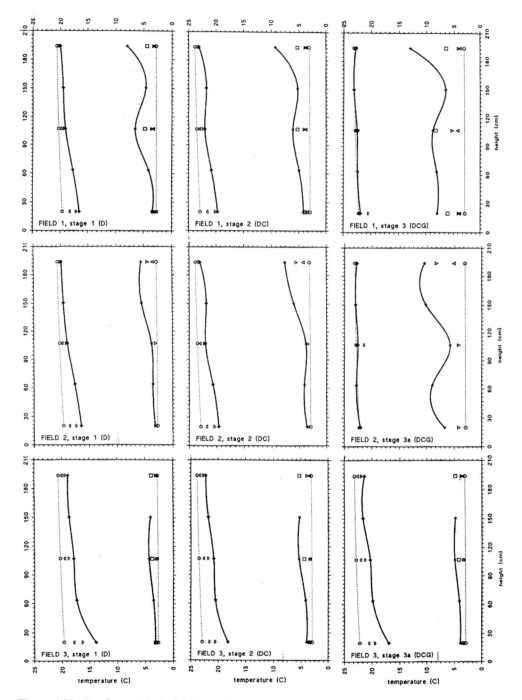

Figure 6.20. Roof assembly 2 (field 1), 3 (field 2) and 4 (field 3), temperatures at the inside surface and the underlay.

Calculated thermal transmittance of the three roofs touched 0.25 W/(m² · K). The apparent thermal transmittances as measured equalled (local heat flow rate, divided by the temperature difference between hot and cold box):

Roof assembly, interface	Apparent thermal transmittance, W/(m² · K)								
	Step 1			Step 2			Step 3		
	Under	Middle	Top	Under	Middle	Top	Under	Middle	Top
1. Underlay	0.084	0.27	0.073	0.087	0.26	0.068	0.73	0.54	0.11
Inside surface	0.46	0.16	0.090	0.46	0.19	0.11	0.04	0.015	0.015
2. Underlay	0.095	0.35	0.65	0.59	0.37	0.59	0.89	1.03	0.97
Inside surface	0.50	0.17	0.05	0.48	0.18	0.05	0.04	0.015	0.005
3. Underlay	0.04	0.25	0.40	0.1	0.28	0.44	0.10	0.24	0.35
Inside surface	0.57	0.27	0.18	0.46	0.25	0.14	0.51	0.25	0.20

These results again prove air looping and exfiltration kill thermal transmittance as an assembly property. What is left is an average heat flow rate at the inside surface for a 1 °C temperature difference between the environment at both sides. The very low apparent thermal transmittances at underlay level suggest wind washing also intervened.

Roof assembly 5 and 6

Roof assembly 5 and 6 were part of a test-building project and as such subjected to an indoor temperature, typical for residential buildings and the vagaries of a moderate outdoor climate. Both gable saddle roofs had the insulation layer on top of the rafters, representing a roof type, commonly called 'sarking roof'. Assembly 5 was in to out composed of:

- Gypsum board inside lining
- Air space, some 5 cm thick
- Thermal insulation, XPS-boards, $d = 10$ cm
- Laths and battens
- Concrete tiles

Assembly 6 differed from 5 as it had an air and vapour retarder below the rafters and a vapour permeable underlay with the overlaps taped, see Figure 6.21.

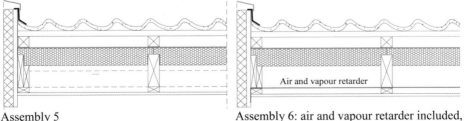

Assembly 5

Assembly 6: air and vapour retarder included, underlay with taped overlaps

Figure 6.21. Roof assembly 5 and 6.

6.4 Performance evaluation

The two roofs got thermocouples in all interfaces and heat flow meters glued against the inside lining at the eaves, in the middle and at the ridge. The winter average thermal resistances logged equalled:

R_{app} $m^2 K/W$	Roof assembly 5 Without air/vapour barrier No underlay Lazy workmanship		Roof assembly 6 With air/vapour barrier Underlay with taped overlaps Good workmanship	
	SW pitch	NE pitch	SW pitch	NE pitch
Ridge	2.43	2.82	3.57	3.28
Middle	2.49	3.12	3.33	3.29
Eave	1.31	2.05	2.69	2.54
Mean	2.09	2.74	3.24	3.10
Overall mean	2.37		3.17	

Once more, the on-site data for assembly 5 underline conduction based concepts such as thermal resistance and thermal transmittance lose content once wind washing and air looping intervene. The differences between both assemblies – 6 thermally better than 5 – endorse the importance of good air and wind tightness for insulation quality. Figure 6.22 pictures the impact of the airflow pattern developing in the two assemblies in terms of Nusselt number versus wind speed normal to the SW pitch, Nusselt being defined as:

$$Nu = \frac{R_{app,measured}}{\sum_{j=1}^{n} R_j}$$

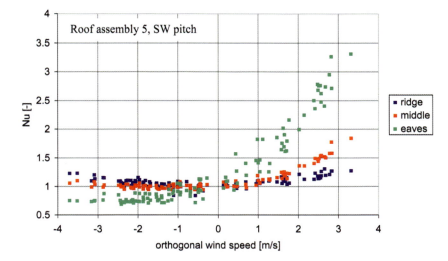

Figure 6.22. Roof assembly 5 and 6, Nusselt versus wind speed normal to the SW pitch.

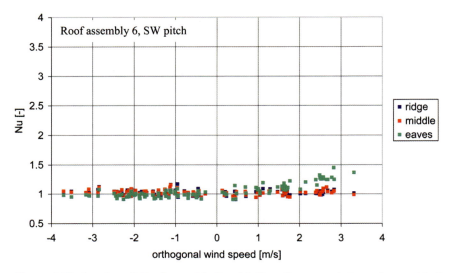

Figure 6.22. (continued) Roof assembly 5 and 6, Nusselt versus wind speed normal to the SW pitch.

Thermal bridges

Pitched roofs are hardly thermal bridge sensitive. Of course, as mentioned in Table 6.10, the rafters and ribs lift the thermal transmittance somewhat in the case of bay filling insulation. Things change for pitched roofs with steel or concrete supporting structure, see Figure 6.23. Then, only cathedralized ceilings with the insulation upon the support perform well.

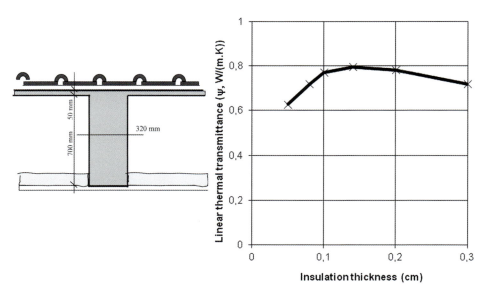

Figure 6.23. Pitched roof, concrete structure, insulation between the concrete rafters. Rafter linear thermal transmittance depending on insulation thickness.

6.4.2.3 Transient response

Most cathedral ceilings are light-weight structures. Therefore, even with excellent air tightness high temperature damping and good admittance remains a dream, as Table 6.11 proves.

Table 6.11. Cathedralized ceilings: transient properties.

Assembly: Gypsum board Thermal insulation (MW) Cavity Underlay	U W/(m² · K)	Temperature damping		Dynamic thermal resistance		Admittance	
		D_θ –	ϕ_θ h	D_q m² · K/W	ϕ_q h	Ad W/(m² · K)	ϕ_{Ad} h
MW, $d = 6$ cm	0.49	1.6	4	2.2	1	0.69	3
MW, $d = 20$ cm	0.17	4.9	7.8	6.6	3.4	0.74	4.4

Even a 20 cm thick insulation does not approach the component related requirements of a temperature damping passing 15 and an admittance beyond $h_i / 2$ W/(m² · K), at 3.9 W/(m² · K). Nonetheless, this is not a problem as Table 6.12 illustrates by showing the inside temperature during a sunny summer day at the end of a heat wave for a 8.4 m deep and 7.2 m wide inhabited attic space below a saddle roof with slope 40°. The SW pitch contains two dormer windows, each 0.56 m² large with $U_{glass} = 1.1$ W/(m² · K). Without good ventilation and the sky lights solar shaded, attic temperature reaches uncomfortably high values. Good ventilation and some shading however, help a lot.

Table 6.12. Saddle roof with cathedral ceiling: attic temperature during a sunny day at the end of a heat wave.

Assembly Gypsum board 20 cm MW Underlay	U W/(m² · K)	Hardly ventilation No solar shading		Ventilated (2 h⁻¹) Solar shading inside	
		Mean °C	Maximum °C	Mean °C	Maximum °C
With tiles (light red)	0.17	39.3	40.4	25.6	27.2
With slates (dark grey)	0.17	43.4	44.7	26.3	28.2

6.4.2.4 Moisture tolerance

As with low-sloped roofs, rain, building moisture and interstitial condensation are the main moisture sources that demand consideration.

Rain

Wind driven rain and precipitation wet the cover, which should therefore be rainproof. For that to be true, junctions and details must be designed in a way they accomplish the requirement (see design and execution). It is nevertheless better to presume that heavy wind-driven rain and powder snow will seep through the cover, which is why the underlay demands detailing as second drainage plane. So, every upper strip must overlap the one below and have a side lap with the adjacent strip. The underlay has to dewater in the roof gutters while chimneys and dormer windows need a sunken gutter all around at underlay level.

Cover materials such as ceramic and concrete tiles suck water during rain events. In moderate climate summers they dry and wet. In winter however, surface condensation by under cooling and the small vapour pressure gradient with outdoors impedes drying, allowing moisture content in tiles and laths to pass capillary, as was measured for tiles on a NE-oriented, well insulated, airtight cathedral ceiling pitch, see Figure 6.24. Measurements that way also showed insulation quality hardly affects the tile's moisture content. Of course, the cover is colder above a well-insulated pitched roof, which is why they endure more frost/thaw cycles and why ceramic tiles see frost damage risk increase, see Table 6.13. Long lasting wetness also favours moss and algae growth on and between tiles.

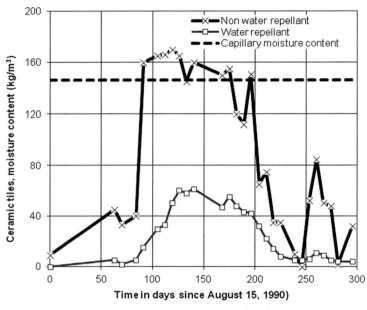

Figure 6.24. Cathedralized ceiling roof, measured moisture content in the tiles.

Table 6.13. Cathedralized ceiling roof, NE looking pitch, frost/thaw cycles measured in a moderate climate region.

Cathedral ceiling roof	Heating season	Frost/thaw cycles	
		Outdoors	In the cover
Airtight, 20 cm MW	86–87, cold winter	37	62
	87–88, mild winter	3	9
	88–89, mild winter	11	16
	90–91, cold winter	39	64 to 70
Air permeable, 8 cm MW	86–87, cold winter	37	69
	87–88, mild winter	3	8
	88–89, mild winter	11	15
Not insulated	86–87, cold winter	37	65
	87–88, mild winter	3	7
	88–89, mild winter	11	10

6.4 Performance evaluation

Building moisture

A new timber supporting structure contains building moisture. Measurements showed this is not a problem. Even an airtight cathedralized NE pitch, thermal transmittance 0.17 W/(m² · K), PE air and vapour retarder below the insulation and a vapour retarding glass fabric reinforced synthetic underlay above, saw building moisture dried within 1 year. Only the first year, some building moisture condensate deposited against the underlay in winter, see Figure 6.25. A really vapour permeable underlay will still accelerate drying.

Figure 6.25. Airtight cathedralized ceiling (assembly 9, see below), building moisture condensing against a vapour retarding perforated glass fibre reinforced synthetic underlay. The grey rectangles in the array below indicate the weeks during the first year when deposit was noted.

Week	1	2	3	4	5	6	7	8	9	10	11	12	13	14	15	16
Up																
Middle																
Down																

Interstitial condensation

In general

If lack of air tightness is a true nuisance in terms of thermal transmittance, the consequences in terms of interstitial condensation are even more serious. While in airtight pitched roofs, the problem is easily neutralized in moderate and cold climates by mounting a leak free vapour retarder at the winter warm side of the insulation, see Table 6.14, in an air permeable pitched roof the absence of leaks is a matter of luck. Indeed, approximately 90% of all pitched roofs lack air tightness.

Ceiling or cathedralized pitch airtight

For the vapour retarding quality needed in moderate climates, see Table 6.14. An insulation which offers the vapour thickness required, needs no additional retarder. As a reminder, in case the insulation requires an additional vapour retarder, the notations E1 and E2 mean:

Class	Boundaries	Examples
E1	$2 \text{ m} \leq [\mu d]_{eq} < 5 \text{ m}$	Bituminous craft paper with overlapping flanges taped
		Gypsum board with aluminium foil finished backside
E2	$5 \text{ m} \leq [\mu d]_{eq} < 25 \text{ m}$	Synthetic foils, $d = 0.2$ mm, mounted with taped overlaps

$[\mu d]_{eq}$ indicates that not only the material but also the way the vapour retarder is mounted defines the effective diffusion thickness. Leaks in fact kill any vapour-retarding quality.

Table 6.14. Vapour retarding quality needed.

Thermal insulation at ceiling level				
Ceiling deck	ICC	Case 1	Case 2	Case 3
Heavy (concrete, prefab structural floor units), no perforations	1	No requirement		
	2/3	No requirement		
	4/5	Analyse case by case		
Timber, insulation not covered, airtight ceiling or perfect air barrier below insulation	1	No requirement		
	2/3	No requir.	E1	E2
	4/5	Analyse case by case		
Timber, insulation covered Airtight ceiling or perfect air barrier below the insulation	1	No requirement		
	2/3	E1	E1	E2
	4/5	Analyse case by case		

Cathedral ceiling	
Necessary for sufficient air tightness	
Insulation filling the bays	**Insulation below ribs or rafters**
Perfectly mounted air retarder at the inside of the insulation. Insulation materials: MW, cellulose	Air-tightened insulation layer consisting of airtight materials such as EPS, XPS, PUR, CG

ICC	Case 1	Case 2	Case 3
1	No requirement		
2/3	No requirement	E1	E2
4/5	Do not apply that type of cathedral ceiling		

As well for the insulation at ceiling level as for cathedral ceilings, case 1, 2 and 3 mean:

	As underlay	As roof cover
Case 1	None	Ceramic tiles, fibre cement slates, corrugated sheets
	Vapour permeable and capillary	All, except shingles
Case 2	None	Ceramic tiles, fibre cement slates, corrugated sheets
	Not capillary bur mounted strip wise	All, except shingles
Case 3	Boarding	Shingles
	Non capillary and continuous	All, except shingles

No vapour retarder is needed when the insulation below offers the vapour thickness required.

6.4 Performance evaluation

	Insulation upon the ribs or rafters
	Joints between the insulating roof deckings or insulation boards air-tightened

ICC	Case 1	Case 2	Case 3
1		No requirement	
2/3	No requirement	E1	E2
4/5	Do not apply that type of cathedral ceiling		

With the insulation upon the ribs or rafters, case 1, 2 and 3 mean:

	Roof cover
Case 1	Ceramic tiles, fibre-cement slates, corrugated fibre-cement sheets
Case 2	Quarry slates, concrete tiles, metallic tiles
Case 3	Shingles

To draft Table 6.14, the measured equivalent diffusion thicknesses of Table 6.15 were used.

Table 6.15. Roof coverings: equivalent diffusion thickness.

Roof cover	Equivalent diffusion thickness m
Roman tiles, single interlock	0.19 ± 0.04
Pan tiles, double interlock	0.27 ± 0.05
Concrete tiles	0.65 ± 0.09
Fibre cement slates	0.87 ± 0.12
Quarry slates	2.08 ± 0.27
Fibre cement corrugated sheets	0.84 ± 0.12
Metallic tiles	1.75 ± 0.22

In- and exfiltration experimentally

Roof assembly 1

Windy weather always under-pressurizing one or more pitches and winter thermal stack inducing exfiltration whenever possible explains the in- and outflow sensitivity of pitched roofs. Testing proved the consequences are annoying. Let us return to assembly 1, see 'thermal transmittance':

- Lathed groove and tongue ceiling
- Air cavity, $d = 20$ mm
- Glass fibre blankets, 54 mm thick, with as vapour retarding layer aluminium coated craft paper. The flanges did not overlap at the ribs
- Air cavity, $d = 90$ mm
- Ceramic Roman tiles

Table 6.16. Assembly 1: interstitial condensation, exfiltrating air as main vapour carrier.

Phase	Condensate deposited on the tiles g/(m² · day)
1. Mainly diffusion, some exfiltration	1.8
2. Leak ∅ 20 mm in the ceiling, overpressure HB 1 Pa	22.3
3. Leak ∅ 20 mm the ceiling, overpressure HB 7 Pa	79.3

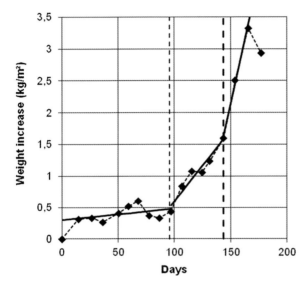

Figure 6.26. Cathedralized ceiling assembly 1: weight increase due to interstitial condensation during test stage 1 to 3.

Cold box climate: dry bulb ≈ 1.5 °C, vapour pressure 580 to 612 Pa. Hot box (HB) climate: dry bulb ≈ 20 °C, vapour pressure 1573 to 1335 Pa. Each six weeks, test conditions changed. During phase 1 the ceiling was kept intact and overpressure in the hot box limited to 1.5 Pa. In stage 2, a hole with diameter 20 mm was drilled across the lathed ceiling reducing the overpressure in the hot box to 1 Pa but increasing the air permeability of the ceiling. In stage 3, overpressure in the hot box went to 7 Pa. At each stage, the assembly suffered from interstitial condensation, yet the amounts condensing increased substantially between stage 1 and 3, see Table 6.16 and Figure 6.26. With the ceiling leak and 7 Pa overpressure, 71 times as much condensate deposited than in stage 1, when diffusion was the main driving force.

The conclusion is clear. Avoiding severe interstitial condensation presumes exfiltration is excluded as much as possible. Or, air tightness is a primary requirement also from a moisture tolerance point of view. To realize that, an air barrier is needed somewhere in the assembly.

In- and ex-filtration, wind washing and air looping experimentally

Besides exfiltration, cathedral ceilings also see infiltration, wind washing, and air looping developing. This combined action makes interstitial condensation complex to predict, as following experimental data show.

6.4 Performance evaluation

<u>Roof assemblies 2, 3 and 4</u>

Let us return to the assemblies 2, 3 and 4, already discussed under 'thermal transmittance' Assembly 2 in to out was composed of:

- Gypsum board lining (quite vapour permeable but strongly airtight)
- Cable cavity, $d = 22$ mm
- Glass fibre, $d = 120$ mm, density 17 kg/m³ (bays between ribs fully filled)
- Fibre cement cellulose underlay (FCC), $d = 3.2$ mm, strip width 120 cm. FCC combines good vapour permeability ($\mu d = 0.14$–0.25 m) with moderate capillarity ($A = 0.011$ kg/(m²·s$^{0.5}$)). The overlaps between strips make the underlay quite air permeable
- Laths and battens
- Glazed ceramic tiles with double interlock

Assembly 3 was identical, though without underlay. Assembly 4 had as underlay a perforated, glass fabric reinforced synthetic foil with a strip width 120 cm. The conditions in the hot and cold box are listed in Table 6.17.

During stage 1, air pressure difference remained zero. Stage 2 saw an overpressure of 18 Pa in the hot box. In stage 3, it went back to 2 Pa, but the gypsum board lining received two 4 mm wide crosscuts normal to the slope. Stage 1 only saw a limited weight increase of the thermal insulation, the underlay and the tiles, see Table 6.17. During stage 2, weight increase was definitely augmented in assembly 4, while stage 3 showed the largest increase in 2 and 3. The picture anyhow is more diffuse than with exfiltration only. As assembly 4 had quite a vapour retarding underlay, air pressure difference impacted more than crosscutting the internal lining. For the assemblies 2 and 3 the inverse was true. Further, air-looping makes condensation height dependent with a maximum in the upper 1/3 for limited looping and more deposit in the lower part for strong looping (Figure 6.27). Wind washing in turn curbs condensation deposit.

Table 6.17. Cathedralized ceiling assemblies 2 to 4: interstitial condensation.

Ass.	Stage	θ_i	θ_e	Δp_{ie}	Stage	Total weight increase g/m²		
		°C	°C	Pa		Insulation	Underlay	Tiles
2	1	20.2	2.8	626	1. Mainly diffusion	0	240	0
	2	23.4	3.0	502	Overpressure HB 1.5 Pa	0	260	35
	3	22.7	2.8	722	2. Overpressure HB 18 Pa	20	450	40
3	1	20.2	2.8	626	3. Overpressure HB 2 Pa	0		0
	2	23.4	3.0	502	Two 4 mm wide crosscuts in the internal lining	0		70
	3	22.7	2.8	722		25		0¹|0|200
4	1	20.2	2.8	626		25	20	0
	2	23.4	3.0	502		20	160	35
	3	22.7	2.8	722		25	70	0

[1] from left to right: weight increase of the under, middle and upper 1/3rd of the tiles

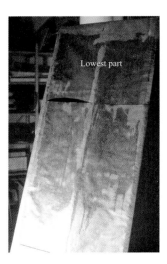

Figure 6.27. Cathedralized ceiling assembly 4, condensation on the underlay. Largest deposit at the lowest part of the underlay.

Roof assemblies 5 to 17

Meanwhile, a 7 years programme ran, monitoring interstitial condensation in 13 NE looking mono-pitch cathedral ceiling assemblies, subjected to a residential indoor climate and a moderate outdoor climate. Surface of the 1.8 wide and 3.6 long pitches: 6.48 m². Indoor temperature: 16 to 18 °C, vapour release: 430 to 1000 g/day, limited ventilation rate. For all assemblies tested, see Figure 6.28.

In a first phase an airtight, well-insulated assembly, which figured as the reference, and an assembly with a vented cavity between insulation and underlay were tested:

Assembly 5, figuring as the reference	Vented assembly 6
Gypsum board internal lining	Gypsum board internal lining
Cable cavity, 22 mm	Insulation: glass fibre blankets, 80 mm
Air and vapour retarder: PE-foil of 0.2 mm	Vented cavity, 120 mm wide
Insulation: mineral wool, 200 mm	
FCC underlay (capillary, vapour permeable)	FCC underlay (capillary, vapour permeable)
Laths and battens, ceramic tiles	Laths and battens, ceramic tiles

The second phase saw the gypsum board internal lining exchanged for a lathed groove and tongue timber ceiling. Before phase 3 started, a micro-perforated, glass fabric reinforced synthetic foil replaced the fibre cement cellulose (FCC) underlay. In phase 4 we removed the vented assembly and installed assemblies 12 and 13, both with an identical cross section as the reference with the synthetic foil underlay, however without air and vapour retarder. Assembly 12 received a lathed groove and tongue timber ceiling, assembly 13 a gypsum board internal lining. Phase 5 finally encompassed four assemblies, all with identical section as the reference.

6.4 Performance evaluation

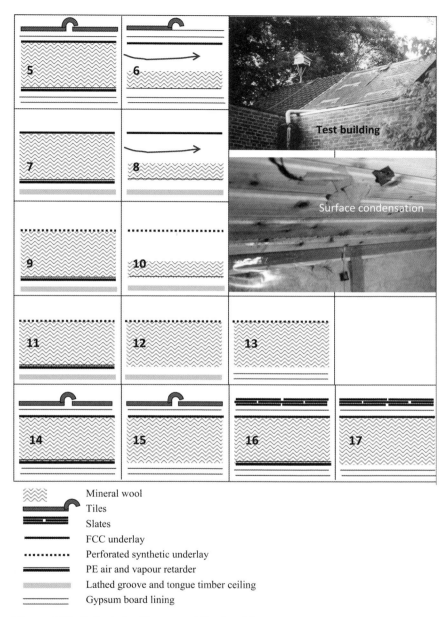

Figure 6.28. Cathedral ceiling assemblies 5 to 16.

Anyhow, that reference, now assembly 14, received its FCC underlay back. Assembly 15 had, like the reference, a tiled cover but missed an air and vapour retarder. Assembly 16 received slates as cover and had an air and vapour retarder below the insulation. Assembly 17 also had a slated cover but missed the air and vapour retarder. In assembly 14 and 16, the indoor finish and the air and vapour retarder was perforated at the top.

Each phase took one or two years. Table 6.18 and the Figures 6.29, 6.30 and 6.31 condense the results. Air tightness is of decisive importance (compare assembly 9 with 12), but not always sufficient. A diffusion resistance, adapted to the nature of the underlay and the indoor climate class, also matters (see assembly 13). Neither does one get, contrary to what one-dimensional calculation packages assume, a homogeneous distribution of the deposit all over the roof surface. Air looping, wind washing and exfiltration at local leakages all intervene. Of course homogeneity increases when the outflow is more equally distributed all over the roof surface (see the assemblies 12, 13, 15 and 17).

Table 6.18. Cathedralized ceiling assemblies: measured condensation response.

	Assembly	**Vapour pressure excess indoors, interstitial condensation**
Assembly 5, W/(m²·K) Tiles	Airtight but vapour permeable internal lining, air and vapour retarder, bays fully filled with mineral wool, capillary, vapour permeable underlay	$\Delta p_{ie} = 348 - 9.9\,\theta_e$ No condensation. Moisture content in the underlay remains hygroscopic
Assembly 6 W/(m²·K) Tiles	Quite airtight but vapour permeable internal lining, no air and vapour retarder, bays partially filled, vented cavity below capillary, vapour permeable underlay	$\Delta p_{ie} = 348 - 9.9\,\theta_e$ No condensation. Moisture content in the underlay remains hygroscopic
Assembly 7, W/(m²·K) Tiles	See assembly 5, now with air permeable but more vapour retarding lathed ceiling	$\Delta p_{ie} = 469 - 12.5\,\theta_e$ No condensation. Moisture content in the underlay hygroscopic
Assembly 8 W/(m²·K) Tiles	See assembly 6, now with air permeable but more vapour retarding lathed ceiling	$\Delta p_{ie} = 469 - 12.5\,\theta_e$ No condensation. Moisture content in the underlay hygroscopic
Assembly 9 W/(m²·K) Tiles	See assembly 5, now with air permeable but more vapour retarding lathed ceiling and a non-capillary, vapour retarding underlay	$\Delta p_{ie} = 824 - 39\,\theta_e$ A little deposit the first winter (building moisture). Later no condensation, Figure 6.29
Assembly 10 W/(m²·K) Tiles	See assembly 6, now with air permeable but more vapour retarding lathed ceiling and a non-capillary, vapour retarding underlay	$\Delta p_{ie} = 824 - 39\,\theta_e$ Severe droplet formation on the underlay, most in the middle, least at the top, Figure 6.29. Surface condensation during cold spells.
Assembly 11 W/(m²·K) Tiles	See assembly 5, however with air permeable but more vapour retarding lathed ceiling and a non-capillary, vapour retarding underlay	$\Delta p_{ie} = 724 - 28.4\,\theta_e$ No condensation, Figure 6.30. Final moisture ratio in the insulation 0.8% kg/kg
Assembly 12 W/(m²·K) Tiles	As assembly 11 but without air and vapour retarder	$\Delta p_{ie} = 724 - 28.4\,\theta_e$ Severe, uniformly distributed droplet formation on the underlay, Figure 6.30. Final moisture ratio in the insulation: 16.6% kg/kg

6.4 Performance evaluation

Table 6.18. (continued)

	Assembly	Vapour pressure excess indoors, interstitial condensation
Assembly 13 0.16 W/(m²·K) Tiles	As assembly 11, however with quite airtight but vapour permeable internal lining but without air and vapour retarder	$\Delta p_{ie} = 724 - 28.4\,\theta_e$ Severe, uniformly distributed droplet formation on the underlay, Figure 6.30. Final moisture ratio in the insulation: 17.7% kg/kg
Assembly 14 0.16 W/(m²·K) Tiles	As assembly 5, the internal lining and the air and vapour retarder perforated at the top	$\Delta p_{ie} = 524 + 5.2\,\theta_e$ Condensation at the top, close to the leak, Figure 6.31
Assembly 15 0.16 W/(m²·K) Tiles	As assembly 5 but without air and vapour retarder	$\Delta p_{ie} = 524 + 5.2\,\theta_e$ Despite the capillary underlay severe condensation. Drips on the insulation. Most condensate at the top, see Figure 6.31
Assembly 16 0.16 W/(m²·K) Slates	As assembly 5, now slated and the internal lining and air/vapour retarder perforated at the top	$\Delta p_{ie} = 524 + 5.2\,\theta_e$ Condensation at the top, close to the leak, Figure 6.31
Assembly 17 0.16 W/(m²·K) Slates	As assembly 5 now slated and without air and vapour retarder	$\Delta p_{ie} = 524 + 5.2\,\theta_e$ Despite the capillary underlay severe condensation. Drips on the insulation. Most condensate at the top, see Figure 6.31

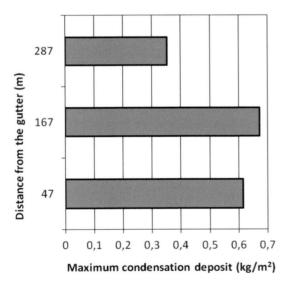

Figure 6.29. Assembly 10.
On the left the maximum condensation deposit measured on the underlay.
In the time array below the grey rectangles indicates the weeks when interstitial condensation was noted.
Dark grey indicates the week(s) with maximum deposit.

Week →	1	2	3	4	5	6	7	8	9	10	11	12	13	14	15	16	39	40	41	42	43	44	45	46	47	48
Up																										
Middle																										
Down																										

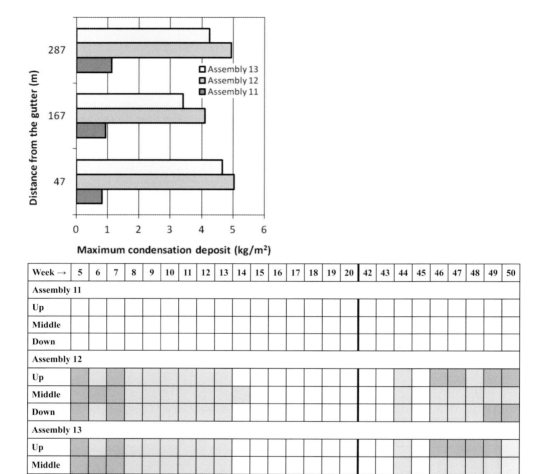

Figure 6.30. Assemblies 11, 12 and 13. On top the maximum condensation deposit measured on the underlay. In the time array below the grey rectangles indicate the weeks when interstitial condensation was noted. Dark grey indicates the week(s) with maximum deposit.

6.4 Performance evaluation

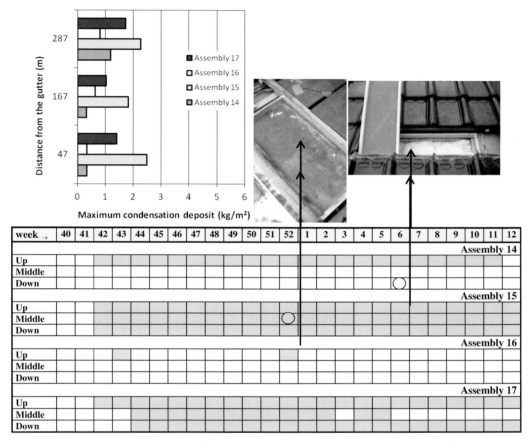

Figure 6.31. Assemblies 14, 15, 16 and 17. Above on the left the maximum condensation deposit measured on the underlay. Above on the right a picture of the FCC underlay as it looked with and without condensate deposited. In the time array below, the grey rectangles indicate the weeks when interstitial condensation was noted. Dark grey indicates the week(s) with maximum deposit.

<u>Roof assemblies 18 to 23</u>

The tests on the thirteen mono-pitch roofs were followed by a hot box/cold box experiment on six cathedralized ceiling pitches to better understand the impact of the underlay's vapour tightness. For the assemblies tested, see Table 6.19. To simulate current practice, the six roofs were not especially air-tightened. Testing ran with a winter mean climate in the cold box, comfort conditions in the hot box and a variable air pressure difference across the roofs with related exfiltrating airflow. Table 6.20 condenses the results and makes a comparison with the predictions according to Glaser, which only considers vapour diffusion. The calculations deviate substantially from what was measured. Remarkable but true, underlay vapour permeance and capillarity appear more important than the diffusion resistance of the layers at the inside of the thermal insulation.

Table 6.19. Cathedralized ceiling roofs, the six assemblies.

	Underlay	$(\mu d)_{eq}$[1] m	Cap?	Insulation thickness Vapour retarder	Internal lining
18	Bituminous foil	32–12	No	14 cm MW, kraftpaper	Gypsum board
19	Spunbonded foil	0.01	No	14 cm MW, kraftpaper	Gypsum board
20	Hydrodiode	1.8–0.01	Yes	14 cm MW, kraftpaper	Gypsum board
21	Bituminous foil	32–12	No	20 cm MW, kraftpaper	Painted gypsum board
22	Spunbonded foil	0.02	No	20 cm MW, kraftpaper	Gypsum board
23	Bituminous foil	32–12	No	12 cm EPS	Gypsum board

[1] Lowest value measured at high mean relative humidity (92%), highest value measured at low mean relative humidity (27%)

Table 6.20. Results of a Glaser calculation, measured results.

	Glaser, condensation (Y/No), where? Evaluation (accepts/do not accept.)	Exfiltrating airflow $m^3/(m^2 \cdot s)$			Underlay Dimensionless temperature factor			Condensation on underlay $g/(m^2 \cdot day)$			Dripping condensate		
18	Y/underlay/accept.	0	$3\cdot10^{-4}$	$7\cdot10^{-4}$	0.07	0.09	0.12	3	24	22	N	Y	Y
19	No	0	$3\cdot10^{-4}$	$7\cdot10^{-4}$	0.10	0.12	0.20	0	0	0	N	N	N
20	No	0	$3\cdot10^{-4}$	$7\cdot10^{-4}$	0.09	0.09	0.15	3	0	5	N	N	N
21	Y/underlay/accept.	0	$4\cdot10^{-4}$	$9\cdot10^{-4}$	0.09	0.14	0.17	1	19	34	N	Y	Y
22	No	0	$4\cdot10^{-4}$	$9\cdot10^{-4}$	0.06	0.11	0.15	0	18	7	N	Y	Y
23	Y/underlay/accept.	0	$4\cdot10^{-4}$	$9\cdot10^{-4}$	0.18	0.27	0.34	4	4	13	Y	Y	Y

Roof assemblies 24 to 31

All results until now concern mono-pitch roofs with limited dimensions, which is why a test building campaign on eight saddle roofs, each spanning 7.2 m with a 45° slope had to confirm the findings (Figure 6.32). The first series included two compact cathedralized ceiling roofs, the one with (assembly 24) and the other without PE air and vapour retarder below the insulation (assembly 25), and two vented ones, one again with (assembly 26) and the other without PE air and vapour retarder (assembly 27). The vented cavity under the underlay was 5 cm wide. All four had a vapour permeable underlay and received a 19 cm thick mineral wool insulation, of which 14 cm was between air and vapour barrier if present, and 5 cm filled the service space above the gypsum board internal lining. The building was an indoor climate class 3 environment, 21 °C in winter for a mean vapour pressure excess touching 600 Pa.

Globally the vented assemblies 26 and 27 performed worse. Wind speed and wind direction had a much larger impact in terms of a higher effective thermal transmittance. Moisture ratio by weight in the rafters close to the underlay climbed higher, while assembly 27 without air and vapour retarder saw the highest condensation rate deposited underneath the underlay, though the values were not problematic. Both compact assemblies 24 and 25 performed excellently: very stable thermal transmittance close to the conduction-related value expected and moisture ratio by weight in the rafters never passing 20% kg/kg. The assembly 25 without air and vapour retarder anyhow saw somewhat more condensate deposited at the underlay than the assembly 24 with.

6.4 Performance evaluation

Figure 6.32. Test building used for testing the two times four saddle roofs (the four in the parallelogram).

The second series included again two compact (assembly 28 and 29) and two vented cathedralized ceiling roofs (assembly 30 and 31), one with (27, 30) and the other without PE air and vapour retarder below the insulation. Lighter glass fibre insulation replaced the mineral wool, used in the assemblies 24 to 27, while the vented assembly 30 with air and vapour retarder received a vapour retarding underlay. The four gypsum board internal linings were also painted. Two years of testing revealed that the lighter glass fibre facilitated wind washing, with less stable thermal transmittances as the main consequence. Wind washing and the low air permeance of the painted gypsum board also produced a positive effect. In none of the four roofs, interstitial condensation happened.

Conclusions

The 31 test roof results showed how to construct cathedralized ceiling pitched roofs with good moisture tolerance in moderate and cold climates. A set of five performance requirements took over from the two traditional recommendations – vapour retarder below the thermal insulation and a vented cavity between insulation and underlay:

1. Make the combination internal lining and air retarding layer, if any, as airtight as needed. How airtight, depends on the indoor climate class and vapour permeance of the underlay.
2. Avoid local leakages in the lining, such as cracks, open overlaps and perforations by cables.
3. Apply a vapour permeable, or, better, a capillary and vapour permeable underlay. Aim with a non-capillary one for a diffusion thickness 0.02 m. For a capillary one, 0.15 m is a good upper threshold.
4. Assure the internal lining and air retarding layer, if any, together touch the vapour retarding quality, listed in Table 6.14.
5. Exclude any air looping possibility around and wind washing in and under the insulation.

The Table 6.21 recommendations connect values to the requirements 1, 2 and 3. The table holds for moderate climates. For a given indoor climate class and diffusion thickness of the underlay it shows the maximal allowable air permeance of the cathedralized ceiling assembly. The layers at the inside of the insulation are assumed to have an E1 vapour retarding quality, though this is not utterly important. In fact, with less air tightness then required, vapour resistance hardly plays a role. The criterion used is that after a cold week condensation on an initially droplet-free non-capillary underlay should not exceed 100 g/m². Above, dripping risk tends to one. At Uccle, the following climate data characterize this cold week:

θ_e °C	ϕ_e %	q_r W/m²	h_{ce} W/(m²·K)	v_w m/s
−2.5	95	−30	17	3.8

with v_w mean wind speed measured in open field at a height of 10 m and q_r resulting combined long and short wave radiation on a pitch facing north with a 30° slope.

Table 6.21. Cathedralized ceilings, maximum allowable assembly air permeance to avoid unacceptable interstitial condensation (value in m³/(m²·s·Pa) for a 5 Pa air overpressure indoors).

Indoor climate class	Maximal air permeance K_a at a 5 Pa overpressure indoors ($K_a = a \cdot 5^{b-1}$) m³/(m²·s·Pa)	
	Vapour retarding underlay	Vapour permeable underlay
	$[\mu d]_{eq} = 2$ m	$[\mu d]_{eq} = 0.02$ m
1	$0.6 \cdot 10^{-4}$	$1.1 \cdot 10^{-4}$
2	$0.2 \cdot 10^{-4}$	$0.3 \cdot 10^{-4}$
3	$0.1 \cdot 10^{-4}$ (1)	$0.2 \cdot 10^{-4}$
4/5	The best here is to opt for solutions with the insulation on an air-tightened boarding	

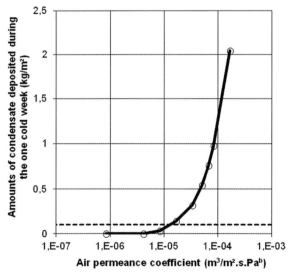

Figure 6.33. Calculation result as the basis of Table 6.21. The cathedralized ceiling has a thermal transmittance 0.4 W/(m²·K) and a vapour permeable underlay of μd = 0.02 m.

For other slopes, radiation is:

$$q_r = 30\left[0.55\cos(s_s) + 1.1\,e_L - 0.6\,a_S - 0.1\right] \tag{6.2}$$

6.4 Performance evaluation

e_L being long wave emissivity and a_S short wave absorptivity of the roof cover. The calculations that generated the table were based on a 5 Pa air overpressure indoors (for a calculation example, see Figure 6.33).

In practice, requirement 5 means: opt for compact cathedralized ceilings, use as thermal insulation mineral wool boards with high enough density (18 kg/m^3 is a lower threshold) or cellulose, fill the bays between ribs or rafters carefully and completely, tape the overlaps between all underlay strips, fill the service cavity above the inside lining with mineral wool once all wire guiding rods are mounted.

6.4.2.5 Thermal bridges

The junctions with other envelope parts are delicate: attic floor/front face; attic floor/back face; attic floor/side walls; sloped roof/low-sloped roof; concrete gutters, etc. Figure 6.34 shows wrong and right examples. Table 6.22 lists the linear thermal transmittances and lowest temperature ratios for the details 1 and 2. Surface film resistance indoors for the linear thermal transmittances was 0.13 m$^2 \cdot$ K/W. For the temperature ratio, we used 0.25 m$^2 \cdot$ K/W.

Table 6.22. Pitched roofs, linear thermal transmittance (ψ) and temperature ratio (f_{hi}) for details 1 and 2 of Figure 6.34 for the insulation thicknesses given in the columns 2 and 3.

Detail	Façade cm	Attic floor cm	ψ W/(m $\cdot$ K)	f_{hi} –
1. Side wall, insulation on the attic floor, no thermal cut	10	10	0.35	0.58
		14	0.38	0.59
		20	0.40	0.60
		30	0.39	0.62
2. Side wall, insulation on the attic floor, correct thermal cut (cellular concrete in the inside leaf)	10	10	–0.09	0.81
		14	–0.07	0.83
		20	–0.06	0.85
		30	–0.07	0.86

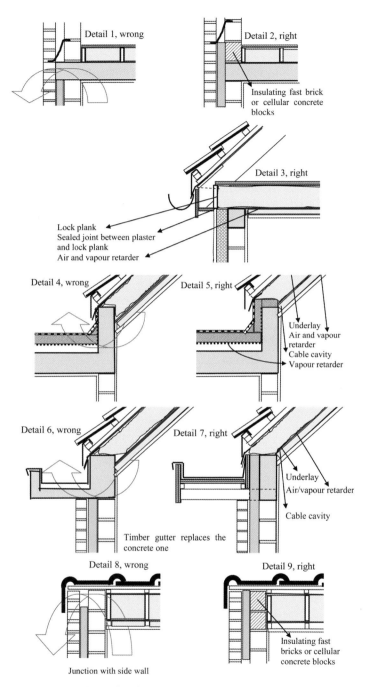

Figure 6.34. Above, insulation at ceiling level: (1, 2) junction with side wall (concrete floor), (3) junction with side wall (timber floor). Below, cathedralized ceiling: (4, 5) junction with flat roof, (6, 7) gutters, (8, 9) junction with side wall.

6.4 Performance evaluation

6.4.3 Building physics: acoustics

Pitched roofs are mainly light-weight structures. A good noise reduction consequently demands application of the rules for composite assemblies: two leafs with different stiffness at large enough distance, a soft, sound absorbing material in between and air tightness as perfect as achievable. In cathedralized ceilings cover and underlay form the one, the inside finish the other leaf, while the mineral wool, glass fibre or cellulose insulation figures as sound absorbing layer.

Table 6.23 lists sound transmission losses measured on cathedral ceilings that are moisture tolerant and have thermal transmittances between 0.2 and 0.6 W/(m² · K). The reference is a tiled non-insulated pitch with a fibre cement cellulose underlay (FCC) but without inside lining. The table proves thermal insulation hardly upgrades sound insulation. Installing an airtight inside lining, however, is a true step forwards. The heavier the lining, the larger the step. Resilient fastening demands mounting the inside lining on a separate system of lightweight steel sections. And, with an airtight lining, more insulation adds some extra sound insulation. The best assembly has a sound transmission loss of some 50 dB.

Insulation and a quite heavy, airtight, resilient inside lining also guarantees good impact noise insulation, which is important to mask impinging rain.

Table 6.23. Cathedralized ceiling roofs, sound transmission loss: measured values.

Roof assembly					R_w
Roof cover	Underlay	Insulation	Air/vapour retarder	Inside lining	(x = 32 dB)
Tiles	FCC	None	None	None	x
		5 cm MW	None	None	x + 2 dB
			PE-foil	Lathed ceiling	x + 9 dB
				Gypsum board	x + 9 dB
				Gypsum board + plaster	x + 14 dB
				Gypsum board, resiliently fastened	x + 22 dB
		15 cm MW		Gypsum board, resiliently fastened	x + 26 dB
	Insulated roof decking, EPS d = 10 cm			Gypsum board + plaster	x + 18 dB

6.4.4 Durability

Despite the important temperature differences noted between winter and summer in moderate climates, hygrothermal deformations hardly damage scutillate coverings. Table 6.24 lists the measured temperatures on a NE looking tiled and the calculated ones for a SW looking slated cathedralized ceiling pitch. Thanks to the scullillate character, the deformations of each separate cover element do not add.

Insulated roof decking elements (particleboard or plywood/thermal insulation/particle board or plywood) react differently. In winter, they bend concave, in summer convex. Also between day and night, they bend, which produces the cracking noises, people having rooms in the attic, complain about.

Table 6.24. Cathedralized ceiling roofs, cover temperatures.

Assembly (in to out): Gypsum board Insulation: 20 cm MW Underlay	Temperature, cold winter day °C		Temperature, hot summer day °C		$\Delta\theta_{year}$ °C
	Min.	Max.	Min.	Max.	
Calculated (Leuven, latitude 51° North)					
Slates, SW, 40° slope	−21.7	−	−	77.9	**99.6**
Measured (Leuven, latitude 51° North)					
Tiles, NE, 40° slope	−9.9			57.4	**67.3**

In winter ceramic tiles go beyond capillary wet (see former Figure 6.24). This makes frost a primary risk. The tile industry never tires in claiming that venting the air space under the tiles prevents damage. For that purpose, they sell expensive ventilating tiles. But testing proved a tiled deck vents well without such tiles, although this does not prevent the tiles from becoming very wet and cold. So, venting hardly decreases the freezing load. Only frost resistant tiles bring relief. Moisture ratio in laths and battens also reaches values in winter that are high enough to initiate mould and rot, which is why they are treated for application outdoors.

Hygric movement can induce such important bulging of large format fibre cement slates, that fixing brackets come out. Accompanying tensile stresses may even break the slates. Finally, synthetic underlay foils must stay protected by the roof cover as they age by UV.

6.4.5 Fire safety

A 30′ fire resistance is guaranteed by combining non-combustible insulation materials with an inside lining of fire safety class A, while the air and vapour retarding foil is best self-extinguishing. If more severe requirements apply, a concrete supporting structure with non-combustible boarding in cellular concrete or prefabricated structural floor units should replace timber.

6.4.6 Maintenance

Gutters demand yearly cleaning. Between tiles and slates, moss may grow. In severe cases, only mechanical removal helps.

6.5 Design and execution

6.5.1 Roof assemblies

Basis for a correct choice is attic space use. If projected as storage or buffer space, the logic option is insulation at ceiling level. Going for a cathedralized ceiling in such cases makes no sense as it only enlarges the protected volume, i.e. the volume enclosed by the thermal insulation. In addition, when the junction between outer walls and pitched roof lacks air-tightness, the attic will stay so well vented that heat loss by a tempered storey below may remain quite high. Combining a cathedralized ceiling solution with insulation at ceiling level is also not logical. Inhabited attics of course demand cathedralized ceilings, many times in combination with some ceiling level insulation.

6.5.1.1 Attic as storage and buffer space

As was said, the solution is insulation at ceiling level, see Figure 6.35. Two remarks: the best solution with a timber deck is fully filling the bays between the joists with mineral wool, glass fibre, or cellulose, while design and execution must respect the requirements in terms of air and vapour retarding quality discussed above.

Figure 6.35. Insulation at ceiling level.

6.5.1.2 Inhabited attic

The three ways to construct cathedralized ceilings are: (1) insulation layer upon the supporting structure, (2) insulation filling the bays between ribs or rafters, (3) Insulation layer under the ribs or rafters. One of course can also prefabricate the cathedral ceiling pitches.

Insulation layer upon ribs or rafters

Variants are:

- Insulated roof decking elements (Figure 6.36). These consist of sandwiches 'particle board or plywood/thermal insulation/particle board or plywood'. An alternative are particleboard or plywood sheets with ribs nailed upon and insulation (PUR or PIR) in between these ribs. Roof decking elements replace ribs, insulation, underlay, and battens. The critical points are air tightness of the joints between elements and the junction details at valley gutters, roof strings, ridges, gutters, roof windows, chimneys, etc. The elements as such are airtight.

144   6 Pitched roofs

- Self-bearing insulation boards (Figure 6.37). The most common are PIR, EPS and XPS boards. They replace the underlay. Rain, wind, and air-tightness are realized by groove and tongue jointing and covering the boards with a vapour permeable, though rain and airtight foil. A horizontal support rib at the gutters prevents the boards from sliding along the rafters. The insulation boards as such are quite airtight.
- Outside insulation (Figure 6.37). In such case, the rafters get a plywood or OSB boarding first with a glued air and vapour retarding polymer bitumen on top. The ribs with the height demanded by the insulation thickness are then nailed across the retarder into the boarding, after which one fills the bays between ribs with semi-dense mineral wool or glass fibre boards. An underlay finally covers the whole, after which the battens and laths are nailed and the cover laid. The solution deserves recommendation in indoor climate class 4 and 5 premises where lack of air and vapour tightness can end in dramatic moisture problems, a risk that must be minimized.

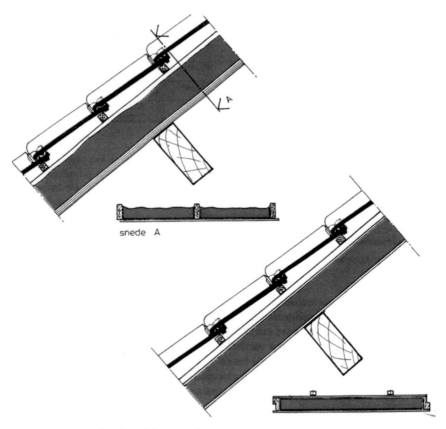

Figure 6.36. Insulated roof-decking elements.

6.5 Design and execution 145

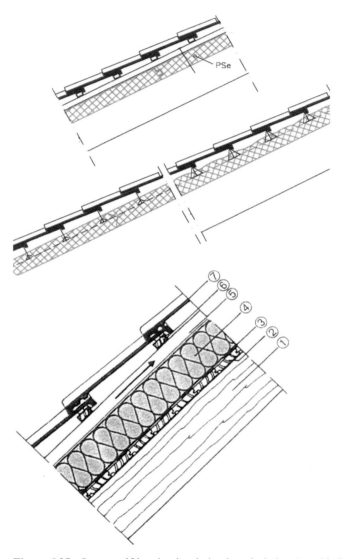

Figure 6.37. On top self-bearing insulation boards, below 'outside insulation'.

Insulation filling the bays between ribs or rafters

The best solution consist of filling the bays between rafters or ribs up to the underlay completely with mineral wool or glass fibre boards, the last with density above 18 kg/m³. Mounting starts with cutting the boards 1 cm wider than the distance between rafters or ribs. That way they tighten the bays nicely without gaps (Figure 6.38). Ribs and rafter height follows the insulation thickness, even if that is more than what is needed structurally. A thermal transmittance 0.2 W/(m² · K) for example demands a height of 20 cm. For larger insulation thicknesses, ribs or rafters are furnished with engineered cross sections, minimizing the thermal bridge effect that way. An alternative is to fill the bays with blown cellulose.

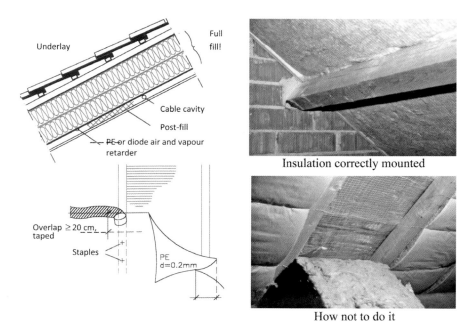

Figure 6.38. Insulation filling the bays between ribs or rafters.

Once the insulation is mounted, a PE or diode air and vapour retarder is stapled against the ribs or rafters with the overlaps carefully taped and the junctions with trusses, partition walls and outer walls sealed. The term diode applies for a foil whose vapour resistance drops with increasing relative humidity. Under the air and vapour retarder a service cavity is left, which afterwards is filled with mineral wool or glass fibreboards. Also the overlaps between the vapour permeable spun bonded or capillary, vapour permeable FCC-underlay strips are taped, while special attention must go to wind-tightening the pitch/gutter junction.

Insulation layer below the ribs or rafters

A possibility is using composite panels 'insulation/gypsum board', which are screwed against the ribs or rafters. Thereafter, the joints between the panels and the joints at trusses, partition walls and outer walls are sealed so air-tightness is guaranteed (Figure 6.39), followed by finishing.

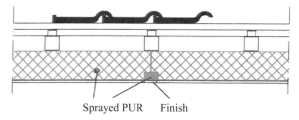

Figure 6.39. Insulation layer below the ribs or rafters.

6.5 Design and execution

6.5.2 Roof details

The ultimate performance of a pitched roof depends on the quality of the details. Each time, the following questions must be answered:

1. Are they correct from a structural integrity point of view?
2. Do they keep the rain proofing function of the cover and the underlay intact?
3. Do they not favour wind washing?
4. Do they respect the thermal insulation continuity?
5. Do they not endanger air tightness?
6. Do they not figure as a diffusion leak?
7. Do they not degrade fire safety?
8. Are they buildable?

Specific details include gutters, ridges, roof strings, valley gutters, roof windows, chimneys, junctions with rising outer walls and junctions between pitched and low-sloped. Figure 6.40 collects some examples for cathedralized ceilings with the insulation filling the bays between ribs or rafters.

It is up to the reader to see how these details solve the questions posed here. The colour code may help (red for the thermal insulation, bleu for the air and vapour retarder, yellow for the drainage planes and wind barriers). The examples given of course are not exhaustive. Each insulation system demands specific solutions.

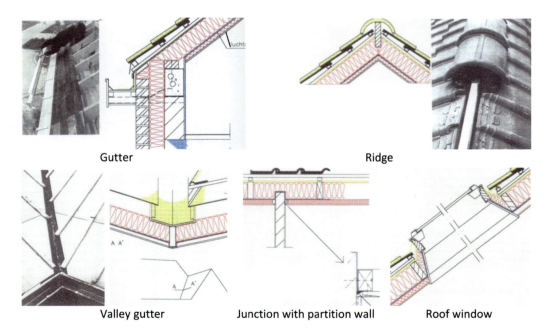

Figure 6.40. Cathedralized ceiling, the insulation filling the bays between ribs or rafters: roof details.

Figure 6.41 shows some details developed on demand for cathedral ceilings with self-bearing insulation boards upon the ribs or rafters. A weak point with these and insulated roof decking elements is that only a few manufacturers offer solutions for these details in terms of appendages and instructions how to execute them. This resulted in experimenting at the building site, sometimes with detrimental results in terms of leakages. Prefabricating complete pitched roofs is a way out, but only if all details are correctly solved. Modular construction anyhow is a precondition to turn such step into a success.

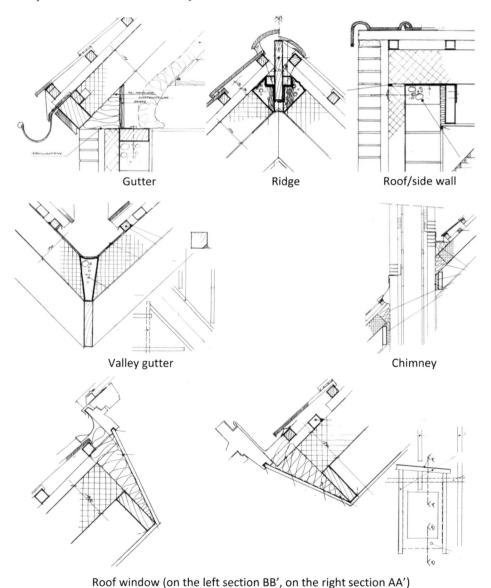

Figure 6.41. Cathedralized ceiling with self-bearing insulation boards upon the ribs or tiles, roof details.

6.6 References and literature

[6.1] Hens, H. (1971). Cursus constructie van gebouwen, teksten en documentatie (in Dutch).

[6.2] Technische Voorlichting 130 (1980). *Plaatsing van asfaltleien op daken*. WTCB, 35 pp. (in Dutch).

[6.3] Tveit, A. (1987). *Air flow and moisture accumulation in slanting wooden roofs*. CIB-W40, Borås meeting.

[6.4] Hens, H. (1987). *Buitenwandoplossingen voor de residentiële bouw: hellende daken*. Nationaal Programma RD Energie, 89 pp. (in Dutch).

[6.5] Technische Voorlichting 175 (1989). *Daken met pannen in gebakken aarde, opbouw, uitvoering*. WTCB, 48 pp. (in Dutch).

[6.6] Fraunhofer Institut für Bauphysik (1990). Kolloquium 'Schrägdach', 12 Juli 1990 (in German).

[6.7] Hens, H. (1991). *Pitched Roofs, Heat-Air-Moisture transport in tiled and slated roofs with the thermal insulation at rafter level*. Laboratorium Bouwfysica, report 90/23, 99 pp.

[6.8] Janssens, A. (1991). *The influence of airflows on the hygro-thermal behaviour of sloped insulated roofs*. Report IEA-Annex 24 Hamtie, 16 pp. + Figs. + Tabs.

[6.9] NEN 6707 (1991). *Bevestiging van dakbedekkingen*. NNI (in Dutch).

[6.10] TI-KVIV (1991). *Studiedag Fysisch gedrag van hellende daken, nieuwste inzichten*. Antwerpen, December 1991 (in Dutch).

[6.11] Hens, H. (1992). *Luft-Winddichtigkeit von geneigten Dächern, wie sie sich wirklich verhalten*. Bauphysik 14, Heft 6, pp. 161–174 (in German).

[6.12] Brueren, H. (1992). *Hoe berekenen we de dakpan verankering?* DIM, Sept., pp. 29–33 (in Dutch).

[6.13] Hens, H., Meert, E. (1992). *Hellende daken*. WTCB-Tijdschrift, Lente (in Dutch).

[6.14] Technische Voorlichting 186 (1992). *Daken met tegelpannen, opbouw en uitvoering*. WTCB, 64 pp. (in Dutch).

[6.15] Janssens, A., Hens, H., Silberstein, A., Boulant, J. (1992). *The influence of underroof systems on the hygro-thermal behaviour of sloped insulated roofs*. Proceedings of the Thermal Performance of the Exterior Envelopes of Buildings V, Clearwater Beach, Florida.

[6.16] Hens, H. (1993). *Pitched Roofs 2, Heat-Air-Moisture transport in tiled and slated roofs with the thermal insulation at rafter level, Effects of air barrier and capillary underlay*. Laboratorium Bouwfysica, report 92/4, 38 pp. + Figs. + Tabs.

[6.17] Thue, J. (1993). *Report on the Norwegian National Construction: lightweight roofs-wooden, sloped, ventilated*. IEA-Annex 24, Report T4-N-93/03.

[6.18] Hens, H., Silberstein, A. (1993). *Pitched Roofs: Effects on Hygric Behaviour of Type of Underlay, Type of Insulation and Vapour Permeability of the Internal Lining*. IEA-Annex 24, Report T4-B-93/01, 16 pp.

[6.19] Janssens, C., Lecompte, J., Meert, E. (1993). *Golfplaten van vezelcement, deel 1: regen- en winddichtheid*. WTCB-Tijdschrift, Winter (in Dutch).

[6.20] NPR 6708 (1993). *Bevestiging van de dakbedekkingen*. NNI (in Dutch).

[6.21] Brueren, H. (1993). *Onderdakfolies vaak onjuist toegepast*. Bouwwereld 25, Dec., pp. 30–32 (in Dutch).

[6.22] Janssens, C., Lecompte, J., Meert, E. (1994). *Golfplaten van vezelcement, deel 2: warmte-isolatie en luchtdichtheid*. WTCB-Tijdschrift, Lente, pp. 30–40 (in Dutch).

[6.23] Rose, W. (1994). *Attic Ventilation in the US: Regulation and research, Part I (1938–1978)*. Building Research Council, University of Illinois, 11 pp.

[6.24] Rose, W. (1994). *Heat Flux, Sheating-Applied Batt Insulation and Loose Fill Cathedral*.

[6.25] Ceilings, Winter 1993 Data, Building Research Council, University of Illinois, 6 pp.

[6.26] Rose, W. (1994). *A Review of Field Thermal Performance of Loose-Fill Insulation in Residential Attics and Cathedral Ceilings*. Building Research Council, University of Illinois.

[6.27] FVB (1994). *Bouwmethodes, 11. Thermische isolatie-3. Hellende daken*. Brussel (in Dutch).

[6.28] Technische Voorlichting 195 (1995). *Daken met natuurleien, opbouw en uitvoering*. WTCB, 52 pp. (in Dutch).

[6.29] Hens, H. (1995). *Pitched Roofs, an Air Permeable Construction*. Baufachtagung 'Gute Luft, wenig Energie'.

[6.30] Künzel, H. M. (1996). *Feuchtesichere Altbausanierung mit Neuartiger Dampfbremse*. Bundesbaublatt 45, Heft 10, pp. 798–801 (in German).

[6.31] Künzel, H. M. (1996). *Tauwasserschäden in Dach aufgrund von Dampfdiffusion durch angrenzenden Mauerwerk*. wksb, Heft 37, 41, pp. 34–36 (in German).

[6.32] Ministère de la Région Wallonne, Direction de l'Energie (1996). *Isolation thermique de la toiture inclinée*. 20 pp. (in French).

[6.33] Technische Voorlichting 202 (1996). *Daken met betonpannen, opbouw en uitvoering*. WTCB, 84 pp. (in Dutch).

[6.34] Ingelaere, B. (1997). *Luchtgeluidisolatie bij pannen- en leiendaken*. WTCB-Tijdschrift, Zomer, pp. 21–28 (in Dutch).

[6.35] Kosny, J., Petrie, T., Christian, J. (1997). *Thermal Bridges in Roofs Made of Wood and light-Gauge Steel Profiles*. ASHRAE Transactions, Vol. 103, Part 1.

[6.36] Künzel, H. (1997). *Steildächer, die Normvorschriften sind überholt*. Bundesbaublatt 46, Heft 5, pp. 312–316 (in German).

[6.37] Parker, D., Sherwin, J. (1998). *Comparative Summer Attic Thermal Performance of Six Roof Constructions*. ASHRAE Transactions, Vol. 104, Part 2.

[6.38] Rudd, A., Lstiburek, J. (1998). *Vented and Sealed Attics in Hot Climates*. ASHRAE Transactions, Vol. 104, Part 2.

[6.39] Janssens, A. (1998). *Reliable control of interstitial condensation in lightweight roof systems*. Doctoraal proefschrift KU Leuven.

[6.40] Janssens, A. (2001). *Advanced numerical models for hygrothermal research: 2DHV model description*. Moisture analysis and condensation control (Ed.: H. R. Trechsel), ASTM Manual 40, pp. 177–178.

[6.41] Künzel, H. M., Leimer, H.-P. (2001). *Performance of Innovative Vapor Retarders Under Summer Conditions*. ASHRAE Transactions, Vol. 107, Part 1.

[6.42] Riesner, K. (2003). *Natürliche Konvektion in losen Außenwanddämmungen*. Rostocker Berichte aus dem Fachbereich Bauingenieurwesen, Heft 12 (in German).

[6.43] Hens, H., Janssens, A. (2004). *Cathedral ceilings: a test building evaluation in a cool, humid climate*. Proceedings of the Performance of the Exterior Envelopes of Whole Buildings IX, Clearwater Beach, Florida (CD-ROM).

[6.44] Houvenaghel, G., Horta, A., Hens, H. (2004). *The impact of airflow on the hygrothermal behavior of highly insulated pitched roof systems*. Proceedings of the Performance of the Exterior Envelopes of Whole Buildings IX, Clearwater Beach, Florida (CD-ROM).

6.6 References and literature

[6.45] Janssens, A., Dobbels, F. (2005). *Ontwikkeling en toepassing van luchtdichtheidscriteria voor hellende daken*. Bouwfysica, 16^e jaargang, No. 2, pp. 7–17 (in Dutch).

[6.46] Dobbels, F., Janssens, A., Houvenaghel, G., Hens, H., Wouters, P. (2005). *Onderzoeksproject 'Vochtgedrag van daken', eindverslag*. WTCB (in Dutch).

[6.47] Anon. (2005). *ASHRAE Handbook of Fundamentals, Chapter 24*. ASHRAE, Atlanta, Georgia, USA.

[6.48] Uvslokk, S. (2005). *Moisture and temperature conditions in cold lofts and risk on mould growth*. Proceedings of the 7th Symposium on Building Physics in the Nordic Countries, Reykjavik, June 13–15, pp. 7–14.

[6.49] Arfvidsson, J. (2005). *Moisture safety in attics ventilated with outdoor air*. Proceedings of the 7th Symposium on Building Physics in the Nordic Countries, Reykjavik, June 13–15, pp. 149–156.

[6.50] Mattsson, B. (2005). *A sensitivity analysis of assumptions regarding the position of leakages when modelling air infiltration through an attic floor*. Proceedings of the 7th Symposium on Building Physics in the Nordic Countries, Reykjavik, June 13–15, pp. 388–395.

[6.51] Desseye, C., Bednar, T. (2005). *Increased thermal losses caused by ventilation through compact pitched roof construction*. Proceedings of the 7th Symposium on Building Physics in the Nordic Countries, Reykjavik, June 13–15, pp. 881–887.

[6.52] Desseye, C., Bednar, T. (2006). *Wind induced thermal losses through compact pitched roof construction, test roof measurements, simulation model and validation*. Research in Building Physics and Building Engineering (Eds.: P. Fazio, H. Ge, J. Rao, G. Desmarais), pp. 459–464.

[6.53] Hens, H., Vaes, F., Janssens, A., Houvenaghel, G. (2007). *A Flight over a Roof Landscape: Impact of 40 Years of Roof Research on Roof Practices in Belgium*. Proceedings of Buildings X, Clearwater Beach, December 2–7 (CD-Rom).

[6.54] Schumacher, C. J., Reeves, E. (2007). *Field Performance of an Unvented Cathedral Ceiling (UCC) in Vancouver*. Proceedings of Buildings X, Clearwater Beach, December 2–7 (CD-Rom).

[6.55] Desseye, C., Bednar, T. (2008). *Wind induced airflow through lightweight pitched roof constructions: test roof element, measurements and validation*. Proceedings of 8th Symposium on Building Physics in the Nordic Countries, Copenhagen, June 16–18, pp. 425–432.

[6.56] Kalamees, T., Kurnitski, J. (2008). *Moisture convection performance of wall and attic floor joint*. Proceedings of 8th Symposium on Building Physics in the Nordic Countries, Copenhagen, June 16–18, pp. 777–784.

[6.57] Essah, E., Adu, E., Sanders, C., Baker, P., Galbraith, P., Graham, H., Kalagasides, A. (2008). *Simulating the energy benefits and reduction in condensation formation that is obtained from houses with cold pitched roofs*. Proceedings of 8th Symposium on Building Physics in the Nordic Countries, Copenhagen, June 16–18, pp. 793–800.

[6.58] Anon. (2009). *ASHRAE Handbook of Fundamentals, Chapter 24*. ASHRAE, Atlanta, Georgia, USA.

[6.59] Roels, S., Deurinck, M. (2010). *Laboratory and in situ evaluation of reflective foils in pitched roofs*. Proceedings of CESBP 1, Cracow, September 13–15, pp. 223–230.

[6.60] Hagentoft, C.-E., Kalagasidis, A. S. (2010). *Mold Growth Control in Cold Attics through Adaptive Ventilation: Validation by Field Measurements*. Proceedings of Buildings XI, Clearwater Beach, December 5–9 (CD-Rom).

[6.61]   Kurkinen, K., Hagentoft, E. (2011). *Application of risk assessment technique to attics*. Proceedings of 9th Symposium on Building Physics in the Nordic Countries, Tampere, May 29 – June 2, pp. 239–248.

[6.62]   Anon. (2011). *ASHRAE Handbook of HVAC-Application, Chapter 43*. ASHRAE, Atlanta, Georgia, USA.

7 Sheet-metal roofs

7.1 In general

With sheet-metal roofs the difference between low-sloped and sloped is hardly relevant. 'Sheet-metal' indicates that at least the cover is metal based. The possibilities range from the cover to a whole ventilated or compact roof executed this way. The roof cover can be 'self-bearing' or 'on boarding':

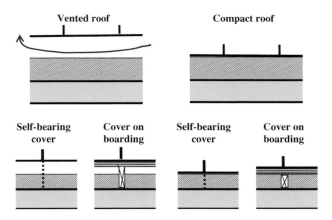

Vented roofs have a ventilated air layer under the cover or boarding. In compact roofs, all layers form one sandwich. Typical assemblies are (top/down):

Vented roof	Compact roof
Metal cover	Metal cover
(boarding)	(boarding)
Vented air layer	
Underlay	(underlay)
Thermal insulation	Thermal insulation
Load bearing deck	Load bearing deck
Inside lining	Inside lining

Vented roofs are site-mounted. Compact roofs may consist of prefabricated elements (Figure 7.1).

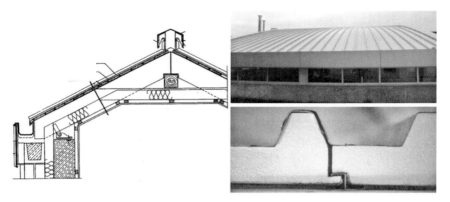

Figure 7.1. Vented metal roof on the left, compact metal roofs on the right (above mounted on site, below consisting of prefabricated panels).

7.2 Metal roof cover

7.2.1 Overview

Lead, cupper, brass, zinc, and aluminium are commonly used. Less common are stainless steel, corten steel, and titanium. For their properties, see Table 7.1. A significant thermal expansion coefficient and quite high stiffness with the exception of lead is typical. Metal roof coverings consequently demand fixing solutions that allow movement. To guarantee that, special mounting and continuity techniques have been developed: tacks, roll caps, and standing seams (Figure 7.2).

Table 7.1. Roof coverings: most important properties of five commonly used metals or alloys.

Metal/alloy	Composition	Density kg/m^3	λ-value W/(m·K)	Thermal expansion coeff. m/(m·K)	Modulus of elasticity MPa	Melting temperature °C
Lead	Pure, or, + 0.03 to 0.06% Cu, + 0.1 to 0.3% Bi	11 336	34.8	$2.9 \cdot 10^{-5}$		327
Copper	+ 0.01/0.03% P (phosphorus)	8 900	360	$1.7 \cdot 10^{-5}$	125 000	1 083
Brass	Copper/zinc alloy	8 750		$1.8 \cdot 10^{-5}$	120 000	1 023
Zinc	+ Cupper, + Titanium	7 140	113	$2.2 \cdot 10^{-5}$		420
Aluminium	+ 1% Mn + traces of Mg	2 700	204–230	$2.3 \cdot 10^{-5}$	67 000	658

7.2 Metal roof cover

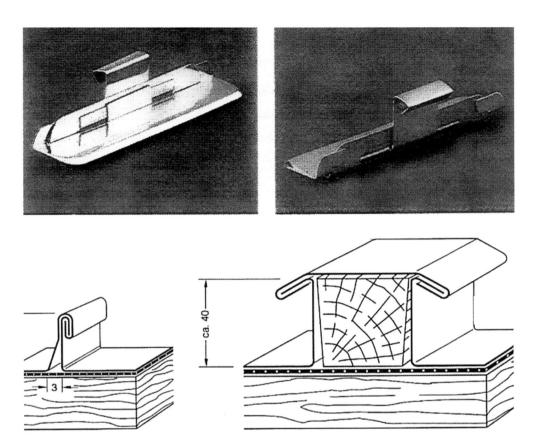

Figure 7.2. On top a tack, below on the left a standing seam and below on the right a roll cap.

7.2.2 Lead

The sheets used have a thickness of 2 to 5 mm. Limiting the dimensions keeps the weight below 100 kg per sheet. Lead shows good corrosion resistance, except if contacting organic acids, lime, or cement. Deformability requires mounting sheets and strips on a boarding, with an interlayer (glass fabric or polyester fabric) between timber and lead. Lead roofs can be compact or vented.

Covering starts at the eaves. The roofer nails or screws the successive sheets at their top while fixing them with tacks at roll caps and bottom overlaps or welts. The copper tacks must be tin-coated. Roll caps guarantee continuity between sheets parallel to the slope, while overlaps or welts do it normal to the slope, see Figure 7.3.

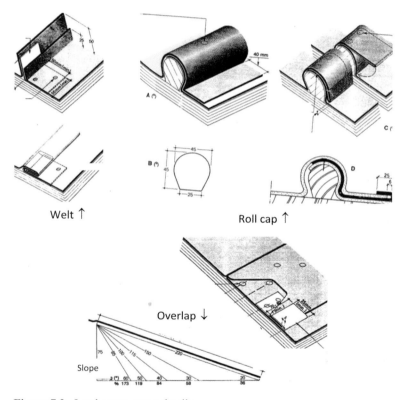

Figure 7.3. Lead cover, some details.

7.2.3 Copper and brass

Copper and brass sheets are 0.6 mm thick. The two have excellent corrosion resistance, but care must be taken when used in combination with other metals. They are much more electro-positive than aluminium and zinc, making run-off water corrosive for these two. Copper and brass are stiff enough to allow self-bearing covers. Of course, mounting on a timber boarding is another option, but an interlayer is then needed.

Self-bearing covers are also fixed on boarding with tacks. Standing seams guarantee continuity parallel to the slope, with the self-bearing covers needing sheet lengths equal to the pitch width. Welts between successive sheets assure continuity when boarding mounted, see Figure 7.4.

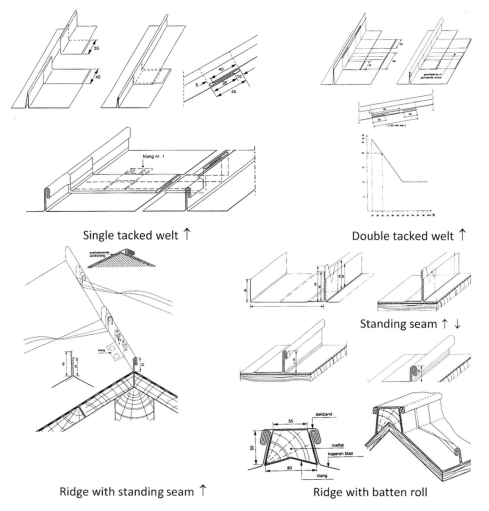

Figure 7.4. Copper cover, some details.

7.2.4 Zinc

Pure zinc is no longer used. 0.6 to 0.8 mm thick zinc/titanium alloy sheets, which show good corrosion resistance at the topside except in the presence of sulphur dioxide (bitumen!), replaced them. The underside, however, may develop quite severe pitting corrosion when subjected to daily relative humidity swings from low to high or drying/condensation cycles. That is why in the last years sheets with protective layer at the underside have replaced naked zinc/titanium. Take care with copper and brass, see above. The sheets are stiff enough to allow self-bearing covers, though boarding solutions are also used. For an example: see Figure 7.5.

Figure 7.5. Zinc cover.

7.2.5 Aluminium

Pure aluminium is too soft to use as roof cover, which is why the 1.2 mm thick sheets mostly consist of an Al-Mn-Mg-alloy with acceptable corrosion resistance. Even better is to laminate the sheets at both sides with another aluminium alloy. Alclad is produced that way. Again, take care with copper and brass. The Al-Mn-Mg sheets have enough stiffness for self-bearing application. Roof elements with corrugated under-sheet, insulation layer (PIR or EPS) and corrugated top sheet are also on the market. Self-bearing covers are fixed using tacks, while parallel to the slope, they demand sheet lengths equal to the pitch width. Standing seams wrapping the tacks guarantee continuity and rain tightness (Figure 7.6).

Figure 7.6. Self-bearing aluminium cover.

7.3 Performance evaluation

We limit the discussion to moisture tolerance, thermal bridging and durability. Structural integrity hardly encompasses metal-specific demands. The roof must be able to bear the design loads without unacceptable sag, while integral metal solutions require bracing to withstand wind without unacceptable deformation.

7.3.1 Moisture tolerance

7.3.1.1 In general

With vented metal roofs, two performance requirements prevailed: thermal transmittance, which had to fulfil the legal requirements if any, and, the air in- and outlet ratio ($p = 100\, A_{\text{vent,in}} / A_{\text{roof}}$) needed to prevent interstitial condensation. For compact roofs, thermal transmittance and vapour barrier quality were the requirements looked at.

However, neither the so-called correct in- and outlet percentages nor an imposed vapour barrier quality kept sheet-metal roofs condensation-free. With compact roofs, building experts had their explanation ready: for a metal roof cover to be vapour tight, the vapour barrier should be perfect. Because tight mounting a foil under insulation seems nearly impossible, preventing interstitial condensation is impossible, which means compact sheet-metal roofs are 'not buildable'. Remarkably, this could not be based on diffusion. Each calculation proved a non-perfectly impervious vapour retarder only gave limited deposit. However when called to analyse damage cases, we always saw abundant condensate, also in compact roofs with vapour barrier. For vented sheet-metal roofs, standardisation bodies sought the answer in a systematic increase of the air in- and outlet percentage, which also failed.

We now know why the diffusion theory was unrealistic and why increased venting was no solution. One overlooked two phenomena for a too long time: under-cooling and lack of airtightness. Their impact on the heat, air, moisture response of metal roofs is quite tremendous.

Under cooling during clear sky nights causes the metal cover to turn colder than the dew point outdoors. Because of that, ventilation in metal roofs becomes a moisture source instead of a drying medium, whereas in compact roofs under-cooling increases the difference between vapour pressure indoors and saturation pressure at the metal cover's underside.

Air flow in and across sheet-metal roofs in turn combines all patterns, among them in- and exfiltration, indoor air washing, wind washing and air looping. The effects resemble those in cavity walls and pitched roofs: thermal transmittance no longer representing the insulation quality, more interstitial condensation in winter, comfort and draught complains, increased sound transmission, higher surface condensation risk on the inside lining, etc. Metal cover solutions only enlarge the effects. To begin with, metals cannot buffer moisture. Condensation then means droplet formation. Further, limited thickness and high thermal conductivity gives metal covers such low thermal resistance, that under-cooling manifests itself equally fast at either the under- or topside, while exfiltrating air flows can hardly warm up the cover.

7.3.1.2 Lack of air tightness

In- and exfiltration

A steady state diffusion/convection model permits estimating the impact of exfiltration on insulation performance and moisture tolerance. Yet, at a sufficiently high outflow, the following equation allows a good estimate of the condensation rate in sheet-metal roofs:

$$g_c = 6.21 \cdot 10^{-6} g_a \left[p_i - p'(\theta_c) \right] \tag{7.1}$$

where g_a is the outflow rate in kg/(m²·s), p_i vapour pressure indoors and $p'(\theta_c)$ saturation pressure underneath the metal cover, a value nearing the one for the equivalent temperature outside. The covering temperature (θ_c) drives the condensation flow rate. Vapour pressure indoors keeps its role: the higher it is, the more severe interstitial condensation, while the deposit is proportional to the air outflow rate. Figure 7.7 portrays calculation results for a roof assembly, in to out composed of a gypsum board, with or without a vapour retarder, 15 cm mineral wool and an aluminium roof cover.

The y-axis gives the condensation deposit noted during a cold week in a moderate climate ($\theta_e = -2.5°$), the x-axis the outflow rate in m³/(m²·h). The deposit first explodes, before passing a maximum at high outflow rate (19 m³/(m²·h)) to drop to zero again when the cover temperature equals the dew point indoors. A vapour barrier gives no relief except when truly airtight.

Outflow at roof level is quasi inevitable. Low-sloped and moderately pitched roofs depressurize when windy, while thermal stack gives overpressure indoors in winter, equal to:

$$p_T = 0.043 (h - h_0)(\theta_i - \theta_e) \tag{7.2}$$

with h the height of the ridge above grade and h_0 the height of the neutral plane indoors. As Table 7.2 shows, metal roofs with the air and vapour retarder stretched under the insulation, lack air-tightness. Only assembly 1 succeeds due to excellent workmanship.

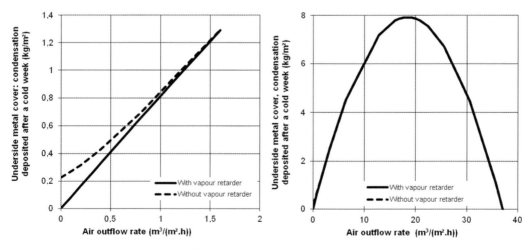

Figure 7.7. Metal roof, condensation deposit after a cold week (–2.5 °C) depending on the outflow rate (indoors 18 °C and climate class 3, relative humidity outdoors 95%, vapour retarder with diffusion thickness 100 m)

7.3 Performance evaluation

Table 7.2. Sheet-metal roofs, air permeance of layers and assemblies.

Layer/assembly	Air permeance		Air outflow for	
	a kg/(m²·s·Pa^b)	$b - 1$ –	$\Delta P_a = 2$ Pa m³/(m²·h)	$\Delta P_a = 10$ Pa m³/(m²·h)
Layers				
Metal cover with standing seams	$\geq 10^{-6}$	−0.36	≥ 0.0047	≥ 0.013
Flange insulation blankets, lazy mounted	$7.5 \cdot 10^{-4}$	−0.15	4.1	15.9
Gypsum board	$2.6 \cdot 10^{-5}$	−0.19	0.137	0.50
PE-foil, perfectly sealed	$4.6 \cdot 10^{-11}$	0	±0	±0
Assemblies				
1. Assembly with stapled PE-foil under the insulation, all overlaps taped	$1.7 \cdot 10^{-6}$	0	0.010	0.05
2. Sandwich aluminium/XPS/aluminium, joints not sealed	$2.1 \cdot 10^{-4}$	0	1.3	6.3

Longitudinal airflow

Outside air

When under-cooling pushes the metal cover temperature in vented roofs below the dew point outdoors, the air passing through will induce condensation. The amounts can cause problems, particularly during somewhat warmer clear sky nights in springtime and autumn. At the same time, apparent thermal transmittance and surface condensation risk at the internal lining will increase drastically, if wind washing accompanies venting.

Inside air

Longitudinal inside airflow is never planned. For it to happen, a slope and leaks in the internal lining at the eaves suffice. When not crossing the air and vapour retarder, nothing happens. If the air can wash the insulation or flow above it, then the consequences are annoying: apparent thermal transmittance, interstitial condensation risk and the amounts deposited, all explode.

Air looping

Air looping prevails when the insulation has an air layer at both sides and is either air permeable or mounted with open joints between the boards. The consequences resemble those in other envelope parts: strong increase in apparent thermal transmittance, higher surface condensation risk, sometimes more interstitial condensation.

7.3.1.3 Under cooling

The metal cover temperature drop due to under cooling follows from following approximate heat balance:

$$q_R + h_{ce}\left(\theta_e - \theta_{se}\right) + \frac{\theta_a - \theta_{se}}{R + \dfrac{1}{h_i}} = 0 \tag{7.3}$$

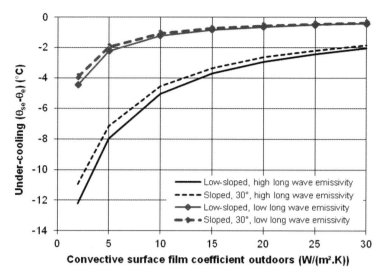

Figure 7.8. Compact sheet-metal roofs with self bearing cover, $R = 5$ m$^2 \cdot$ K/W: under cooling (boundary conditions: $\theta_e = 0$ °C, $\theta_i = 18$ °C).

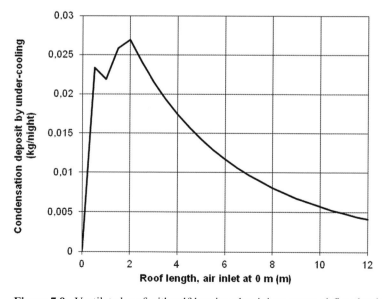

Figure 7.9. Ventilated roof with self-bearing aluminium cover, airflow by thermal stack only (0.3 Pa). Distribution of the under-cooling condensate between air inlet and outlet (boundary conditions: $\theta_e = 0$ °C, $\theta_i = 18$ °C, distance between both 12 m).

7.3 Performance evaluation

with:

$$q_R = 5.67\, e_{Ls}\, F_T \left\{ \theta_e - 21(1-c)\left[F_{ss} + \rho_{L,env}(1-F_{ss})\right] - \theta_{se} \right\} \quad (7.4)$$

In both equations, q_R is the long wave radiant heat flow rate, h_{ce} the convective surface film coefficient outdoors, e_{Ls} the long wave emissivity of the metal cover, F_{ss} the view factor between roof and sky, $\rho_{L,env}$ the albedo of the terrestrial environment, c the cloudiness factor, θ_{se} outside surface temperature of the cover and F_T the temperature factor for radiation between roof and sky. With compact roofs, R is the thermal resistance surface to surface, while θ_a is the (operative) temperature indoors. With vented roofs, R is the thermal resistance of cover and boarding while θ_a is the vented cavity temperature.

Figure 7.8 shows under cooling data for a metal cover with high (0.9) and low (0.2) long wave emissivity. A high long wave emissivity and low convective surface film coefficient increases the effect substantially with a temperature drop between 2 and 12 °C compared to the dry bulb value outdoors. Because of that, important amounts of condensate can deposit underneath the metal cover, most of it close to the air inlets, see Figure 7.9.

7.3.1.4 Two practice and six test building cases

The following two real world cases and test building results confirm the analysis above.

Glass factory

In the early 1990s, a double glass production hall is built. The roof consists of sandwich elements, the cross section in to out consisting of a white enamel aluminium vapour barrier, 50 mm thick EPS-boards with density 15 kg/m³, corrugated aluminium sheet with the EPS glued underneath as cover. Each element was 960 mm wide. The cover sheets overlap each other over one corrugation along the slope while three elements with upper/lower cover sheet overlap are needed to bridge the distance between eaves and ridge. At the vapour barrier level, purpose designed synthetic profiles close the longitudinal joints (Figure 7.10).

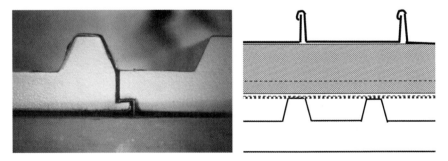

Figure 7.10. Left: cross section of the manufactured sandwich panels, right: the solution for the condensation problem.

The first complaints about dripping moisture occurred shortly after glass production started. Trials to additionally seal the longitudinal joints do not stop the problem. At the end, a hot box/cold box test is ordered on a dummy, consisting of six smaller sandwich elements with the longitudinal joints closed at the inside as done by the manufacturer. The results are confounding:

- Although the elements look airtight, 6.3 m³/(m² · h) air exfiltrates across the roof at 8.5 Pa overpressure in the hot box. A closer look shows the longitudinal and lateral overlaps are quite air leaky. An air layer remains between the EPS and the corrugated cover sheets with the same thickness as the glue lines, while joints with a width up to 6 mm separate the successive 34 cm broad EPS boards. That way, overlaps and air layer are coupled all over the roof surface
- During the test, the water running out of the roof is weighed weekly, giving following relation with the vapour pressure difference over the roof surfaces (test roof area 5.5 m²):

$$g_c = 2.1\,\Delta p - 661 \quad [\text{g/week}] \tag{7.5}$$

- The sheet temperatures logged were 1 to 1.4 °C higher than without exfiltration

After six weeks of testing, the longitudinal and lateral joints in the cover are carefully sealed. This however does not assure air-tightness as we still measure an outflow of 0.43 m³/(m² · h) at 8.5 Pa overpressure in the hot box. The roof went on dripping and the corrugated aluminium sheet temperature remained higher than without exfiltration. This confirmed that only perfect air-tightness excludes dripping moisture. A theoretical alternative could be to have such high air outflow the cover temperature reaches the dew point indoors.

The solution uses the existing roof as load bearing deck for a compact new build-up roof (Figure 7.10).

Dwelling

In a new dwelling water runs out of the roof the first winter (Figure 7.11).

Figure 7.11. Dwelling. Up on the left a view of the metal roof, up on the right traces of moisture leakage, down on the left opening the roof, down on the right leaky vapour retarder along the perimeter.

7.3 Performance evaluation

From the outside to the inside, the assembly looks as follows: self-bearing flexed corrugated aluminium sheets with a span equal to the roof width, cavity, 12 cm thick glass fibre bats, 0.1 mm PE vapour retarder, perforated steel ceiling. When disassembling the roof, it strikes the PE-foil and is far from mounted airtight. It has leaks at all partition walls and along the roof perimeter (Figure 7.11). A blower door test shows an air-tightness of the envelope of $G_a = 587 \, \Delta P_a^{0.605}$ m^3/h , resulting in a ventilation rate of 7.5 ach at 50 Pa for a volume out to out of 964 m^3. A tracer gas measurement confirms the roof takes a substantial part of that leakage, stating once again that lack of air-tightness causes condensation and moisture run out, with air-tightening roof as remedy left.

The way out again was conversion to a compact roof in successive steps of: (1) removal of the corrugated sheets, the insulation and the PE-foil, (2) laying sound absorbing mineral wool boards on the perforated ceiling, (3) mounting a plywood boarding on the existing purlins, (4) gluing an airtight polymer bitumen on these boards over the whole roof surface, including the junctions with the façade walls, (5) nailing 12 cm high ribs across the polymer bitumen and the plywood into above the purlins, (6) filling all bays between the ribs with 12 cm thick, 50 kg/m^3 dense mineral wool, (7) spanning a vapour permeable underlay over the insulation, (8) remounting the corrugated aluminium sheets.

Test building

From 1996 to 2005, test building measurements ran on two vented and four compact zinc roofs, subjected to a moderate climate outdoors (Figure 7.12) and a climate class 3 environment in the building.

Assemblies

Vented roofs	Assembly, in to out
1, more air-tight	Timber purlins as load bearing structure
	Gypsum board ceiling
	0.2 mm PE-foil as airtight layer, overlaps taped
	16 cm mineral wool
	Vapour permeable underlay
	Vented cavity, inlets at the eaves, outlets at the ridge
	Untreated pine boarding, 4″ × 3/4″
	Titan/zinc cover with standing seams
2, less airtight.	As with 1 but with the PE-foil overlaps not taped
Compact roofs	
3, very airtight	Prefab concrete deck
	4 mm polymer bitumen as air-tight layer, melted to the deck with a gas burner
	16 cm thick dense mineral wool boards
	Self- bearing titan/zinc roof cover with standing seams fixed with tacks that stretch through the insulation and are screwed across the airtight layer into the concrete
4, less airtight	As with 3, however instead of 4 mm polymer bitumen a 0.2 mm loose laid PE-foil as airtight layer.
5, less airtight	Structural sheet steel deck
	0.2 mm thick loose laid PE-foil as airtight layer, overlaps taped
	16 cm thick dense mineral wool boards
	Self- bearing titan/zinc roof cover with standing seams fixed with tacks that stay on the mineral fibre boards and are fixed in the structural steel plates with long screws
6, air permeable	As 5, but without PE-air retarder

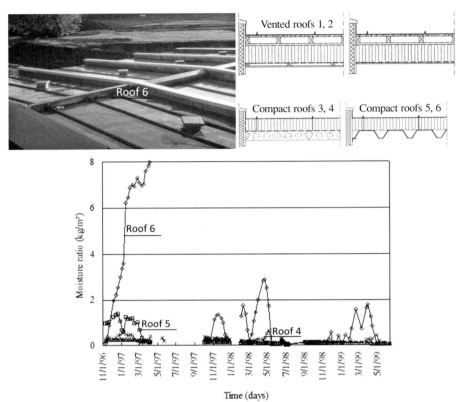

Figure 7.12. Test building: on top the roofs, in the middle the moisture ratio measured in the insulation, below droplet formation.

Vented roof results

Both roofs perform well. A little under-cooling condensate is noted underneath the zinc/titan, while in winter moisture ratio in the pine boarding does not exceed 20% kg/kg and in summer it drops to 5% kg/kg. Relative humidity in the vented cavity remains usually below 100% and the measured thermal transmittance closely matches the design value. The zinc/titan temperature touches a maximum of 72 °C in summer and drops to a minimum of –15 °C in winter.

7.3 Performance evaluation

Differences nevertheless appeared between the more airtight roof 1 and the less airtight roof 2. Temperature and relative humidity are on average higher in the cavity of roof 2. The difference is 0.6 °C, while roof 2 registers 37 days with a relative humidity of 100% in 1997 against 7 days for roof 1. Cover under-cooling in turn is more pronounced in roof 1. Although it is the reference situation for both, the roof 1 cover drops below $(\theta_e - 2)$ °C 84 days out of 100, whereas for roof 2 this is 78 days. $(\theta_e - 8)$ °C and lower is noted 9 days out of 100 for roof 1, but only 4 days for roof 1. And although under-cooling there is less intensive, condensation underneath the zinc during the winter 1996–1997 is the highest in roof 2 (2 weeks compared to none in roof 1), though the amounts remain low The moisture ratio in the boarding also ends up 2 to 3% kg/kg higher in roof 2, while at the end of the winter its purlins contain a little more moisture.

Compact roof results

The differences between the four roofs are more pronounced. The air permeable roof 6 shows the largest moisture deposit in the thermal insulation and abundant droplet formation underneath the zinc/titan during the whole winter 96–97. The very airtight roof 4 on the contrary performs excellently: no moisture deposit in the insulation and hardly any droplet underneath the zinc/titan. Roof 4 and 5 balance in between, with a small moisture deposit in the thermal insulation and less droplet formation than in roof 6, though a little more in roof 4 than in roof 5 (see Figure 7.12). Droplet formation pushes the relative humidity under the cover up to 100%.

Cover under cooling is for all four compact roofs somewhat worse than for the two vented ones. The largest drop below the outside dry bulb temperature reaches 13.5 °C. For roof 3 the zinc/titan temperature drops no less than 2 °C and more for 86 days out of 100 and 8 °C below the dry bulb temperature outside for 14 days out of 100. For roof 4 this is 82 and 7 days, for roof 5 it is 87 and 19 days and for roof 6 it is 86 and 9 days. Between roofs 3 and 4, hardly any difference in thermal transmittance appears and the value noted complies well with the design value.

Conclusions

The two practice and six test building cases underline that vented as well as compact metal roofs may show good moisture tolerance on condition they are acceptably airtight, the insulation layer forms a well-closed layer and has enough density (no bats but semi-heavy mineral wool or glass fibre boards). With vented roofs, a timber boarding operates as a moisture buffer, minimising droplet formation underneath the metal cover by under-cooling. Even at 100% relative humidity in the vented cavity, droplets are rare.

The question remains: how airtight should a sheet-metal roof be to avoid problematic interstitial condensation by air exfiltration? Evaluating droplet run-off risk after a cold week also delivered here the answer in moderate climates. To recall, the weekly mean climate data in the reference climate a 30° sloped roof facing north is subjected to are:

θ_e °C	ϕ_e %	q_r W/m^2	h_{ce} W/(m^2·K)	v_{met} m/s
−2.5	95	−30	17	3.8

Table 7.3 summarizes the results. For an air permeance below $2 \cdot 10^{-5}$ m^3/(m^2·s·Pa) so few condensate deposits underneath the metal roof at a 5 Pa air pressure difference that neither droplet run-off nor metal corrosion demands consideration.

Table 7.3. Metal roofs: mean air permeance needed at 5 Pa air pressure difference to avoid droplet run-off and corrosion.

Indoor climate class	Air permeance $m^3/(m^2 \cdot s \cdot Pa)$
1, 2, 3	$2 \cdot 10^{-5}$
4, 5	Preferred are roofs with the insulation directly on the load bearing deck, which is air and vapour tightened beforehand with deck-bonded 4 mm thick polymer bitumen

Of course, the air permeance left must be area-spread and not the result of local leaks. Besides a correct air permeance, one also needs enough diffusion resistance underneath the insulation to avoid unacceptable interstitial condensation by vapour diffusion, see Table 7.4.

Table 7.4. Metal roofs: vapour retarding quality needed underneath the insulation.

Indoor climate class	Vapour retarder class
1	No requirements
2	$\geq$ E2
3	$\geq$ E2
4, 5	E4

7.3.2 Thermal bridges

Thermal bridging sensitivity is typical for metal constructions. Metals in fact have a high thermal conductivity, so that parts which perforate the enclosure may cause severe heat leakage. But, as for metallic outer wall systems, thin separation layers with low thermal conductivity have a direct positive effect. Good detailing is then based on two rules: (1) avoid metal parts from perforating the thermal insulation layer without thermal cut in between, (2) in case tacks or other fixing elements are screwed in the load bearing substrate below the insulation, put a strong and stiff interfacing block with low thermal conductivity in between (for example neoprene, see Figure 7.13).

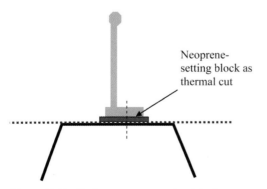

Figure 7.13. Tacks, neoprene setting block.

7.3.3 Durability

We recall that the non-ferrous metals used as roof cover have quite a large thermal expansion coefficient and rather high stiffness, except for lead. If one should fix a metal cover firmly to the substrate, the large temperature swings experienced will generate important forces, which may loosen the couplings and buckle the metal. That is why all fixing solutions have one feature in common: they allow the cover to move (see Figures 7.2 to 7.4).

Moreover, metals are corrosion sensitive. For non-ferrous ones, this is not as bad, although, as Figure 7.14 shows, zinc/titanium without protective layer at the underside may corrode quite severely when used to cover well-insulated assemblies.

 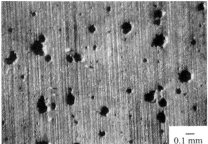

Figure 7.14. Corroded underside of a zinc/titanium cover.

High relative humidity underneath in combination with alternating under-cooling condensation at night and drying during the day cause the pitting corrosion seen. This even happens in airtight compact roof assemblies, insulated with thick mineral wool or glass fibre boards. These in fact contain as much air as their volume allows. For boards stored outdoors, this air is humid, while the fibres also adsorb some hygroscopic moisture. Both together give enough moisture to maintain the daily condensation/drying cycle just mentioned.

Much condensation keeping the zinc/titanium underside permanently wet is safer. Compact and vented roofs also hardly differ in terms of corrosion sensitivity. On the contrary, because nightly under-cooling turns the ventilation air into a moisture source, an airtight, well-insulated compact roof performs better than a correctly built, insulated vented roof.

When in air permeable metal roofs the load bearing structure is timber-based, the winter moisture ratio in the joists, purlins and ribs can exceed the mould threshold of 20% kg/kg. At the same time, indoor climate classes 2 and 3 but surely 4 and 5 environments may induce such abundant interstitial condensation that dripping moisture becomes inevitable!

These three durability aspects once again underline the importance of air-tightness! Furthermore, closed cell insulation materials could help mastering corrosion in airtight compact zinc roofs.

7.4 Design and execution

Clearly, more than for other roof solutions, air-tightness is the main requirement. This is so compelling that only execution of the airtight layer on a deck gives relief. This deck can be light, medium, or heavyweight. Preferentially the air barrier consists of a self-curing material bonded using the gas flame. An appropriate candidate is polymer bitumen.

The legally required thermal transmittance defines the insulation thickness. The layer itself must perfectly link up with the air barrier, by gluing if necessary. Although a perfect air barrier minimizes the risk on unacceptable interstitial condensation, still, depending on the indoor climate class, a certain vapour resistance is needed under the insulation, see Table 7.4. Condensation risk by under cooling in vented roofs is neutralized by (1) mounting the cover on a timber boarding, (2) covering the insulation with a spun bonded underlay foil.

For specific details, we refer to the relevant literature (see references and literature for publications in several languages). As exemplary cases, Figure 7.15 shows how a valley gutter for a vented and compact metal roof is constructed.

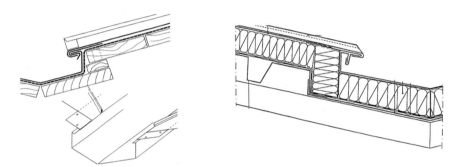

Figure 7.15. Sheet-metal roofs, valley gutters: on the left for a vented roof, on the right for a compact roof.

7.5 References and literature

[7.1] Stark, J., van Zetten, L. (1975). *Dunne staalplaat toegepast bij daken*. Bouwwereld 71, No. 24, pp. 29–33 (in Dutch).

[7.2] Technische voorlichting 169 (1987). *Gebruik van bladlood voor dakbedekkingen en gevelbekledingen*. WTCB, 36 pp. (in Dutch).

[7.3] Frick/Knöll/Neumann/Wienbrenner (1988). *Baukonstruktionslehre, Teil 2*. B. G. Teubner Verlag, Stuttgart, 580 pp. (in German).

[7.4] Wieland, H. (1988). *Korrosionsprobleme in der Profilblech- und Flachdach-Befestigungstechnik*. Praxis und Wissenschaft, 41 pp. (in German).

[7.5] Kleffmann, D. (1989). *Farbaluminium in der Klempnertechnik*. Baumetall, Heft 4, pp. 22–26 (in German).

[7.6] Tietz, A. F. (1989). *Metaal als dakbedekking*. Hoogovens Aluminium, 12 pp. (in Dutch).

[7.7] Alcan Deutschland GmbH (1991). *Unbelüftete wärmegedämmte Metall-Dächer in Klempner-Technik, Ausführung, Besonderheiten, Richtlinien*. 50 pp. (in German).

7.5 References and literature

[7.8] Technische voorlichting 184 (1992). *Daken van koperen bladen en banen*. WTCB, 52 pp. (in Dutch).

[7.9] Kettlitz, J. (Ed.) (1994). *Handboek voor duurzame metalen gevels en daken*. Ten Hagen en Stam BV, Den Haag (in Dutch).

[7.10] Janssens, A., Hens, H. (1996). *Vliet proefgebouw*. 1^e jaarrapport, Laboratorium Bouwfysica, 37 pp. (in Dutch).

[7.11] Janssens, A., Roels, S., De Praetere, W., Hens, H. (1997). *Vliet proefgebouw*. 2^e jaarrapport, Laboratorium Bouwfysica (in Dutch).

[7.12] Kosny, J., Petrie, T., Christian, J. (1997). *Thermal bridges in roofs made of wood and lightgauge steel profiles*. ASHRAE Transactions Vol. 103, Part 1.

[7.13] Hens, H., Janssens, A. (1998). *Bouwfysische aspecten van metalen daken*. Bouwfysica, Vol. 9/1, pp. 16–22 (in Dutch).

[7.14] Zheng, R. (1998). *Zinc roofing and the corrosion of zinc and zinc roofings in the atmosphere*. Predoctoraal rapport, Laboratorium Bouwfysica, 102 pp.

[7.15] Janssens, A., Roels, S., De Praetere, W., Zheng, R., Janssen, H., Hens, H. (1999). *Vliet proefgebouw*. 3^e jaarrapport, Laboratorium Bouwfysica (in Dutch).

[7.16] Zheng, R., Carmeliet, J., Bogaerts, W., Hens, H. (2001). *Method to determine the number and size of samples taken from zinc roofs to analyze pitting corrosion*. Proceedings of the Performance of Exterior Envelopes of Whole Buildings VIII Conference (CD-ROM), Clearwater Beach, Fl., 2–7 December.

[7.17] Zheng, R., Janssens, A., Carmeliet, J., Bogaerts, W., Hens, H. (2003). *An evaluation of highly in sulated cold zinc roofs in a moderate humid climate, Part 1 hygrothermal performance*. Construction and Building Materials, Vol. 18, 49–59.

[7.18] Zheng, R., Janssens, A., Carmeliet, J., Bogaerts, W., Hens, H. (2003). *An evaluation of highly insulated cold zinc roofs in a moderate humid climate, Part 2, corrosion behaviour*. Construction and Building Materials, Vol. 18, 61–71.

[7.19] Janssens, A., Hens, H. (2003). *Interstitial Condensation Due to Air Leakage: A Sensitivity Analysis*. Journal of Thermal Envelope and Building Science, Vol. 27, No. 1, July, pp. 15–29.

[7.20] Hens, H., Jansssens, A., Zheng, R. (2003). *Zinc Roofs: an Evaluation Based on Test House Measurements*. Buildings and Environment, Vol. 38, pp. 795–806.

[7.21] Hens, H., Zheng, R., Janssens, A. (2003). *Does performance based design impacts traditional solutions? Metal roofs as an example*. Research in Building Physics, A. A. Balkema Publishers, 1020 pp.

[7.22] Zheng, R., Carmeliet, J., Hens, H., Janssens, A., Bogaerts, W. (2004). *A Hot Box-Cold Box Investigation of the Corrosion Behavior of Highly Insulated Zinc Roofing Systems*. Journal of Thermal Envelope and Building Science, Vol. 28, No. 1, July, pp. 27–44.

[7.23] Zheng, R., Carmeliet, J., Hens, H., Bogaerts, W. (2003). *The influence of roofing systems on the corrosion of zinc sheeting in highly insulated zinc roofs*. Research in Building Physics, A. A. Balkema Publishers, 1020 pp.

[7.24] Zheng, R. (2004). *Performance of highly insulated zinc roofs in a moderate climate*. Doctoral thesis, Laboratorium Bouwfysica, 207 pp.

[7.25] Haddock, R. M. (2004). *The pitfalls of snow retention on metal roofing*. Proceedings of the Performance of the Exterior Envelopes of Whole Buildings IX, Clearwater Beach, Florida (CD-ROM).

[7.26] Laboratory of Building Physics (2006). *Vliet test building: zinc roofs, actual roofs, corrosion of the zinc sheeting*. New roof assemblies, report 2006/16.

[7.27] Hens, H., Vaes, F., Janssens, A., Houvenaghel, G. (2007). *A Flight over a Roof Landscape: Impact of 40 Years of Roof Research on Roof Practices in Belgium*. Proceedings of Buildings X, Clearwater Beach, December 2–7 (CD-Rom).

[7.28] Zheng, R., Janssen, A., Carmeliet, J., Bogaerts, W., Hens, H. (2010). *Performances of highly insulated compact zinc roofs under a humid-moderate climate – Part I: hygrothermal behaviour*. Journal of Building Physics, Vol. 34, issue 2, pp. 178–191.

[7.29] Zheng, R., Janssen, A., Carmeliet, J., Bogaerts, W., Hens, H. (2011). *Performances of highly insulated compact zinc roofs under a humid moderate climate – Part II: corrosion behaviour*. Journal of Building Physics, Vol. 34, issue 3, pp. 277–293.

[7.30] Anon. (2011). *ASHRAE Handbook of HVAC-Application, Chapter 44*. ASHRAE, Atlanta, Georgia, USA.

8 Windows, outer doors and glass façades

8.1 In general

Transparent parts make the envelope wind and rainproof. The term encompasses glass, frames, and opaque fill-in panels, while the dimensions vary from separate windows to storey-high glass fronts.

In this chapter, we first look at the glass. Then we examine the frames and end with their combination to form windows, outer doors, glass fronts, structural glazing, curtain walls, double skin façades and PV-façades.

8.2 Glass

8.2.1 In general

The most important glass functions are to let in daylight and create contact with outdoors. Building users appreciate both, except in rooms where the activity demands full attention, like meeting rooms and auditoriums, or where the function allows no day lighting. Other performance expectations have emerged in the course of time:

1. Glazed surfaces should not negatively impact thermal comfort
2. The energy demand for heating and cooling per unit area of glass should stay within limits
3. Surface condensation on glass is unwanted
4. Solar gains across glass should not induce overheating
5. Measures to limit solar gains and heat loss should not hamper day lighting
6. Complete darkening must be possible
7. A view to the outside is desirable, looking in is less so
8. Unwanted sound transmission from outside is an annoyance
9. Glazing must be sufficiently strong and stiff
10. It may not degrade fire safety and security against break-in.
11. Maintenance must be easy
12. Durability should largely exceed the warranty period

This package has generated performance requirements for thermal transmittance (1, 2, and 3), solar transmittance (4), visible transmittance (5), sound transmission loss and contact noise insulation (8), fire safety and burglar security (10). In addition the desire for easy maintenance (11) stimulated the development of self-cleaning glass types, though this also includes accessibility. Whether glazing will be durable (12) depends on the technology applied and the care with which the glass panels are manufactured.

Those performance requirements guided glazing type evolution since the 1950s. Of course, the quest for a lower thermal transmittance played an important role. While up to then, single glass was the only option, the offer has since become very diverse. Double glass entered the market first, followed by multiple glass, later-on double glass with low-e coating, and finally low-e double glass filled with better insulating gas than air. In the last two decennia, even low-e, gas-filled multiple glass is available. For each type, manufacturers developed methods to lower solar gains while keeping good visual transmittance. That generated heat absorbing glass and glass with selective short wave reflectivity. The search for better sound transmission loss lead to the introduction of double glass with panes of different thickness, one of them laminated and the cavity filled with an insulating, better noise damping gas than air. Moreover, in search for break-in security and fire safety laminated glass, tempered glass, armoured glass, and swelling foam glass were developed.

8.2.2 Performance evaluation

8.2.2.1 Structural safety

Loads

For vertical glazing wind constitutes the main load. Mounting forces play a subordinate role. Only when wrongly positioned, may these cause problems. For skylights, the weight of the glass, the slope and snow are also factors. All of them bend the glass panels.

Strength

With p the load normal to the surface (N/m²) and L span (m), glass stresses follow from (Pa):

$$\sigma = \frac{M\, y_x}{I}; \quad \sigma_{max, tension} = -\frac{\frac{\alpha\, p\, L^2}{8}\, \frac{d}{2}}{\frac{d^3}{12}}; \quad \sigma_{max, compression} = \frac{\frac{\alpha\, p\, L^2}{8}\, \frac{d}{2}}{\frac{d^3}{12}} \quad (8.1)$$

with M the bending moment, I the moment of inertia per meter run of glass, and d glass pane thickness. α is a form factor and y the ordinate along the glass thickness with the midplane as origin (Figure 8.1, all SI-units).

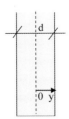

Figure 8.1. Single glazing, thickness, y-ordinate origin at mid-plane.

For glass panes supported at two opposite sides, the span is the distance in between, giving a form factor 1. If instead the pane is supported at four sides, the form factor depends on the ratio between longest (L_2) and shortest side (L_1) with the shortest one seen as span. Knowing the permissible stress, Equation (8.1) directly gives glass pane thickness, for single glass:

$$d = \beta\, L_1 \sqrt{\frac{k}{\sigma_{br}}} \quad d \text{ in m} \quad (8.2)$$

8.2 Glass

where σ_{br} is ultimate strength in Pa, k a safety factor and β a constant equal to $0.866/\alpha$. For opposing supported panes, α is 1. Table 8.1 gives the β/α-values for four-side supported glass. The ultimate strength for either glass type approximates 41.2 MPa, while the safety factor for pressed and tempered glass is 2.5 and for armoured glass, 4.

Table 8.1. Glass pane supported at all four sides, β/α as function of span ratio (L_2/L_1).

L_2/L_1	1.00	1.05	1.10	1.15	1.20	1.25	1.30	1.35	1.40	1.45	1.50
β/α	0.54	0.56	0.58	0.60	0.61	0.63	0.65	0.66	0.67	0.69	0.70
L_2/L_1	1.60	1.70	1.80	1.90	2.00	2.50	3.00	3.50	4.00	4.50	5.00
β/α	0.72	0.74	0.75	0.77	0.78	0.82	0.84	0.86	0.86	0.86	0.86

Equivalent rectangles replace glass panes of any form. For triangular panes supported along the whole perimeter, this rectangle has a same height as the triangle and a length equal to X times the triangle's basis (b), see Figure 8.2. In case of a trapezoidal pane supported along the whole perimeter, that rectangle gets the same height (h) and a length equal to the short side (b_1), increased by X times the difference ($b_2 - b_1$) between short and long side of the trapezoid, see Figure 8.2. For triangle and trapezoid, X quantifies as:

b/h	X
0.50	0.64
0.75	0.55
1.00	0.49
1.25	0.46
1.50	0.43
2.00	0.38

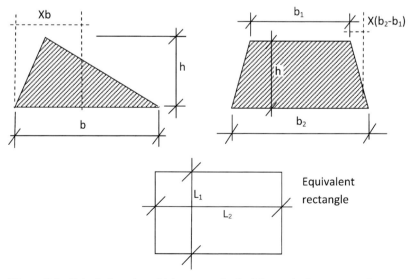

Figure 8.2. Calculating glass thickness, method of the equivalent rectangular pane.

Finally, the equivalent rectangle replacing a circular glass pane supported all around is a square with a side equal to 1.06 times the circle's diameter.

For double-glass with panes of equal thickness, wind load (p) divides equally over both. Multiplying by 1.34 accounts for the temperature and gas pressure changes in the cavity. This gives as design load per pane 0.66 p. If the two have different thickness, wind load distributes proportionally to the third power of thickness:

$$p_1 = \frac{d_1^3}{d_1^3 + d_2^3} \qquad p_1 = \frac{d_2^3}{d_1^3 + d_2^3} \qquad (8.3)$$

Again the results are multiplied by 1.34.

With triple glass with panes of equal thickness each pane takes 1/3 of the wind load (p). Multiplying that load for the two outer panes by 1.34 accounts for temperature and gas pressure changes in the cavities and results in a load distribution 0.5 p / 0.33 p / 0.5 p. With different pane thickness, distribution again occurs proportionally to the third power of thickness, after which the load on the two outer panes is multiplied by 1.34.

Stiffness

Deflection (y) of a glass pane, supported at two opposite sides, follows from:

$$y = \kappa \frac{p \, L^4}{E \, d^3} \qquad (8.4)$$

with E the modulus of elasticity, equal to $7.2 \cdot 10^4$ MPa, and κ a multiplier with value 0.115. Table 8.2 gives that multiplier for glass supported at its four sides. Hardly any requirement limits deflection, though one should avoid values beyond 1/250 of the span for panes supported at two opposite edges or 1/250 of the shortest side for panes supported all around.

Table 8.2. Deflection of a rectangular glass pane supported all around, multiplier κ as function of the ratio between long (L_2) and short side (L_1).

L_2/L_1	1.00	1.20	1.40	1.60	1.80	2.00	2.50	3.00	> 10.0
κ	0.036	0.050	0.062	0.074	0.083	0.090	0.100	0.109	0.115

Lowest resonant frequency

The lowest resonant frequency of a rectangular glass pane is:

$$f = \frac{\pi}{2}\left[1 + \left(\frac{L_1}{L_2}\right)^2\right]\sqrt{\frac{2.4 \cdot 10^6 \, d^2}{L_1^4 \left(1 - \nu^2\right)}} \qquad (8.5)$$

with ν Poisson's coefficient, equal to 0.22. To avoid problems with dynamic wind loads, this frequency must exceed 5 Hz.

8.2.2.2 Building physics: heat, air, moisture

Air tightness

Is not a problem.

Thermal transmittance

Most transparent parts include a frame, glass, and sometimes opaque infill. Glazing used today is double or multi-pane with edge spacer. Also opaque infills may have edge spacers. Together with the frame, spacers make the heat flow three dimensionally, which in theory excludes usage of a thermal transmittance. The following equation is anyhow applied:

$$U_w = \frac{U_{gl} A_{gl} + \psi_{sp,gl} L_{sp,gl} + U_{op} A_{op} + \psi_o L_o + U_{fr} A_{fr}}{A_w} \tag{8.6}$$

with U_{gl} the central thermal transmittance of the glazing, A_{gl} the visible glass area (as seen from outside), $\psi_{sp,gl}$ the linear thermal transmittance of the glass edge spacers, $L_{sp,gl}$ their length, U_{op} the central thermal transmittance of any opaque infill, A_{op} the visible infill area (as seen from outside), ψ_{op} the linear thermal transmittance of the infill edge spacers (equal to 0 without), L_{op} their length, A_{fr} the exterior frame area as seen when orthogonally projected on a plane parallel to the window (allows reading it from the drawings), U_{fr} the equivalent thermal transmittance of the frame and A_w the total area out to out of the window (equal to $A_{gl} + A_{op} + A_{fr}$) (Figure 8.3, all SI-units). The equivalent thermal transmittance of the frame follows from:

$$U_{fr} = \frac{\Phi_{3D,fr}}{A_{fr}} \tag{8.7}$$

with $\Phi_{3D,fr}$ the three-dimensional heat flow across at a 1 °C temperature difference between the environments at both sides.

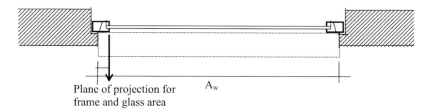

Plane of projection for frame and glass area A_w

Figure 8.3. Window area, as defined when calculating thermal transmittance.

For twofold windows, thermal transmittance becomes:

$$U_w = \frac{1}{\frac{1}{U_{w,1}} - \frac{1}{h_i} + R_c - \frac{1}{h_e} + \frac{1}{U_{w,2}}} \tag{8.8}$$

with $U_{w,1}$ the thermal transmittance of the exterior window, $U_{w,2}$ the thermal transmittance of the interior window and R_c the thermal resistance of the cavity in between. For dual frame windows (two separate glazings on a sane frame), one has (1 the exterior and 2 the interior glazing):

$$U_w = \frac{\left(\dfrac{1}{\dfrac{1}{U_{gl,1} + \dfrac{\psi_{sp,gl1} L_{sp,gl1}}{A_{gl,1}}} - \dfrac{1}{h_i} + R_c - \dfrac{1}{h_e} + \dfrac{1}{U_{gl,2} + \dfrac{\psi_{sp,gl2} L_{sp,gl2}}{A_{gl,2}}}}\right) A_{gl} + U_{fr} A_{fr}}{A_{gl} + A_{fr}}$$

Table 8.3 lists the requirements for the glass central thermal transmittance (U_{gl}) and window thermal transmittance (U_w) as set in several European countries and country regions. Globally, the interval is $1 \leq U_w \leq 4.5$ W/(m² · K) for windows. The main value defining variable is the climate. Portugal for example has a moderately warm climate, while Sweden goes from moderately cold to cold. However legal thresholds for the glass (U_{gl}) to be used are only set by one country and one country region.

Table 8.3. Thermal transmittance of windows and glazing: requirements.

Country (date introduced)	$U_{w,max}$ W/(m² · K)	$U_{gl,max}$ W/(m² · K)
Austria	1.7–1.9	
Belgium, Flanders, 2012	2.2	1.3
2014	1.8	1.1
Denmark (1995)	1.8	
Finland (2003)	1.4	
France (RT 2000)	2.4	
Germany (EnEV 2010)	1.3	1.1
Ireland (2002)	2.2	
Luxemburg	2.0	
Portugal (2006)	3.3–4.3	
Sweden (2006)	1.0/1.1	
UK, England (2002)	2.0–2.2	
UK, Scotland (2002)	1.8	

Single glass

Its (central) thermal transmittance equals:

$$U_{gl} = \frac{1}{\dfrac{1}{7.7} + \dfrac{d_{gl}}{\lambda_{gl}} + \dfrac{1}{25}} = \frac{1}{0.17 + \dfrac{d_{gl}}{\lambda_{gl}}} \qquad (8.10)$$

Glass has a thermal conductivity 0.8 to 1 W/(m · K). If we take 1 W/(m · K) as representative, then a 6 mm glass pane gives a thermal transmittance 5.75 W/(m² · K), i.e. more than 25 times the value that today figures as a life cycle cost optimum for opaque assemblies. Transposing this into gas consumption for space heating by a north oriented pane, gives 35 to 60 m³ per year, with an annual CO_2 release of 62 to 116 kg!

8.2 Glass

Equation (8.10) shows how the surface coefficients fix the single glass thermal transmittance. Glass thickness hardly has an effect. If close to zero, the value is 5.95 W/(m²·K). If 20 mm, it drops to 5.32 W/(m²·K), in absolute terms a gain beyond the step of 0.6 W/(m²·K) but relatively spoken a marginal upgrade, only −11%. Moreover, the value moves up and down with the surface film coefficients. When in stormy weather this coefficient outdoors (h_e) reaches ≈ 100 W/(m²·K), the thermal transmittance nears 7.1 W/(m²·K). If wind-still, the value drops to 4.9 W/(m²·K). Indoors, we see the same. An extreme case is a greenhouse, where all walls have the same temperature and radiant exchanges fade away. The inside surface film coefficient (h_i) then lowers to 3.5 W/(m²·K), which brings the thermal transmittance down to 2.3 W/(m²·K) in wind still weather. A radiator in front takes care of the other extreme. Cold weather and the radiator at 80 °C, pushes it up to ≈ 12.5 W/(m²·K).

Low surface temperatures indoors figure as additional consequence of the high thermal transmittance:

$$\theta_{si} = 0.75\, \theta_e + 0.25\, \theta_i \quad (h_i = 7.7 \text{ W/(m}^2 \cdot \text{K)}, \quad h_e = 25 \text{ W/m}^2 \cdot \text{K)}) \tag{8.11}$$

i.e. a temperature ratio of 0.25. For 21 °C indoors and standard surface film coefficients, ice forms inside each time the outdoor temperature, corrected for under cooling, drops below −7 °C. At 0 °C outdoors, glass temperature does not exceed 5.3 °C, which generates surface condensation each time inside vapour pressure reaches 890 Pa, some 300 Pa above the value outdoors in moderate climates. In an inhabited dwelling with a volume of 500 m³, vapour production can reach 7 to 14 kg per day. If in such case, we want to avoid surface condensation on single glass at 0 °C outdoors and 21 °C indoors, ventilation rate must exceed 0.5 h⁻¹. At 10 °C inside, as noted in unheated sleeping rooms during winter, we need a rate beyond 0.9 h⁻¹.

Also from a thermal comfort point of view, single glass scores badly. At 21 °C indoors and a view factor 0.5 with the glass, comfort complaints will surge each time the outside temperature drops below 0.3 °C. Because of that, already before the energy crisis of 1973, developing glazing solutions with lower thermal transmittance became a priority.

Using reflective foils

Adhering a long wave reflective foil on single glass minimizes the radiant part in the surface film coefficient (Figure 8.4). In fact, the in- and outside values also write as:

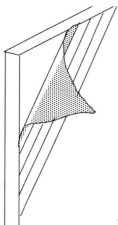

Figure 8.4. Single glass: lower thermal transmittance thanks to reflective foils.

$$h_i \approx 3.2 + 5\, e_L \qquad h_e \approx 21 + 4.5\, e_L$$

with e_L long wave emissivity, transforming the single glass thermal transmittance equation to:

$$U_{gl} = \cfrac{1}{\cfrac{1}{3.2 + 5\, e_L} + d + \cfrac{1}{21 + 4.5\, e_L}} \tag{8.12}$$

which shows that such foils lower thermal transmittance most effectively when adhered inside, see Table 8.4. At long wave emissivity 0.1, the value drops some 72% ($\Delta U_{gl} = -2.4$ W/(m² · K)).

Table 8.4. Thermal transmittance of single glass with a reflective foil at the inside surface.

Long wave emissivity of the foil (e_L)	U_{gl}-value W/(m² · K)	
	Foil inside	Foil outside
0.5	4.52	5.59
0.4	4.20	5.56
0.3	3.86	5.53
0.2	3.52	5.50
0.1	3.16	5.48

However the solution has its disadvantages. A lower inside surface film coefficient brings the surface temperature down, for a long wave emissivity 0.1 to:

$$\theta_{si} = 0.855\, \theta_e + 0.145\, \theta_i \tag{8.13}$$

i.e. a temperature ratio of 0.145. At 21 °C indoors, frost forms on the glass each time temperature outdoors drops below −3.6 °C. At 0 °C outdoors, glass temperature hardly touches 3.0 °C. Does this worsen thermal comfort compared to non-foiled single glass? Not for radiation, as the glass yet acts as a mirror, reflecting the radiant temperature of the half-space in front. Yes for convection, as the air in contact with the foiled glass becomes colder than for non-foiled glass, boosting the air fall and the cold airflow above the floor. The lower inside surface temperature also increases the likelihood of surface condensation, at 0 °C outdoors and 21 °C indoors already for a vapour pressure inside beyond 758 Pa, i.e. 178 Pa above the value outside in a moderate climate. Of course, abundant surface condensation lifts long wave emissivity to that of water, increasing thermal transmittance this way!

Double-glazing

In the 1950s, the search for better thermal comfort gave birth to double glass. Double glass consists of two glass panes separated by a dry air filled cavity and the perimeter hermetically sealed by edge spacers containing desiccant. The air cavity was hermetically closed to avoid condensation at the cavity side of the coldest pane by water vapour diffusing from the environment into the cavity. For the heat transfer across the central part of double glazing, one has (Figure 8.5):

8.2 Glass

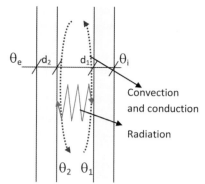

Figure 8.5. Double glass, heat transfer.

Pane 1

$$q = \frac{\theta_i - \theta_1}{0.13 + d_1}$$

Cavity

$$q = \frac{\lambda_a \, \text{Nu} \, (\theta_1 - \theta_2)}{d_c} + 5.67 \, F_T \frac{\theta_1 - \theta_2}{\dfrac{1}{e_1} + \dfrac{1}{e_2} - 1} = \frac{\theta_1 - \theta_2}{R_c} \tag{8.14}$$

Pane 2

$$q = \frac{\theta_2 - \theta_e}{0.04 + d_2}$$

The central thermal transmittance then becomes ($d_1 = d_2 = d$):

$$U_{gl} = \frac{1}{0.17 + 2\,d + R_c} \tag{8.15}$$

The first term in the cavity Equation (8.15) combines conduction and convection, while the second one describes long wave radiation between both bounding surfaces. In case the air cavity is thin enough to block convection, central thermal transmittance gets the values listed in Table 8.5 (for $\theta_e = 0\ °C$ and $\theta_i = 20\ °C$).

Table 8.5. Double-glass with pane thickness 6 mm: central thermal transmittance.

Cavity width	U_{gl}-value	Heat transfer in the cavity	
mm	(0 °C – 20 °C) W/(m²·K)	Conduction + convection %	Radiation %
6	3.31	51	49
8	3.12	44	56
10	3.00	38	62
12	2.91	34	66
15	2.81	29	71

At first glance, the cavity looks miraculous. A width beyond 10 mm halves the central thermal transmittance compared to single glass. This was enough to call double glass 'thermally insulating'. Today, we know better. Double glass is still a heat leak. Of course, compared to single glass, the upgrade makes a difference but compared to the life cycle cost optimum for opaque assemblies, more than 10 times lower, some modesty should be observed. Let it also be noted that the additional 0 °C – 20 °C in Table 8.5, is necessary. In fact, the central thermal transmittance changes with the mean temperature and the temperature difference across.

The lower value results in higher inside surface temperatures, for 6–15–6 mm glass:

$$\theta_{si} = 0.35\,\theta_e + 0.65\,\theta_i \tag{8.16}$$

i.e. a temperature ratio of 0.65. At 21 °C indoors, freezing inside only starts at –39 °C outside. At 0 °C outdoors, the inside surface still reaches 13.7 °C. Vapour pressure indoors must then exceed 1587 Pa before surface condensate deposits. Thermal comfort also improves. For a view factor of 0.5 with the glass, dissatisfaction now requires a temperature below –23 °C outdoors! The higher temperature ratio, however, has a drawback. Exchanging single for double glass in non-insulated or thermal bridge rich buildings may induce mould on the inside surface of opaque envelope spots with a temperature ratio 0.7 or lower.

Multiple glazing

Separating three or more glass panes by dry hermetically sealed air cavities using desiccant containing edge spacers, brings even lower central thermal transmittances into reach, see Table 8.6. Following rule of the thumb allows a quick estimate:

$$U_{gl} = \frac{6}{n} \tag{8.17}$$

n being the number of panes. At first sight, the result looks promising. Quadruple glass nears the thermal transmittance of an unfilled cavity brick wall. Additional gain however drops with the number of panes. A value ≈ 1 W/(m² · K) requires sextuple, a value ≈ 0.6 W/(m² · K) tenfold glass! With panes of equal thickness, total glass thickness becomes $(n-1)\,d_c + n\,d_{gl}$, while total weight increases to 2500 $n\,d_{gl}$. Quadruple glass ($n=4$, $d=6$ mm) with 6 mm wide cavities for example is 42 mm thick and weights 60 kg/m², values that demand such massive frames that applicability becomes questionable. Moreover, solar and visual transmittance decreases while the glass colours greener with increasing number of panes. Also early edge spacer fracture risk goes up. In short, there are enough drawbacks to explain why multiple glazing failed.

Table 8.6. Multiple glass with pane thickness 5 mm: central thermal transmittance.

Cavity width	U_{gl}-value W/(m² · K)	
	Triple glass	Quadruple glass
6	2.30	1.77
8	2.11	1.61
10	1.98	1.49
12	1.90	1.43
15	1.81	1.35

8.2 Glass

Low long wave emissivity double-glazing (low-e double-glass)

Let us return to double glass. According to Table 8.5, long wave radiation accounts for up to 70% of the heat crossing wider cavities. Therefore a logical step towards a lower central thermal transmittance consists of minimizing radiation by covering one or both bounding surfaces with a reflective coating (Figure 8.6). For the result, see Table 8.7. Wider cavities see the central thermal transmittance drop from 30 to 53% compared to double glass with a same cavity width. This supersedes triple glass and nears quadruple glass (Table 8.6), though at half the thickness and a 33 to 50% lower weight!

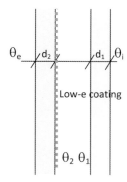

Figure 8.6. Low-e double-glazing.

Table 8.7. Low-e double glass with pane thickness 6 mm: central thermal transmittance.

Cavity width mm	U_{gl}-value W/(m²·K)		
	e_L		
	0.2	0.1	0
6	2.70	2.56	2.40
8	2.38	2.21	2.02
10	2.16	1.96	1.74
12	1.99	1.77	1.52
15	1.80	1.57	1.29

The low-e double glass sold has a central thermal transmittance ≈ 1.8 W/(m²·K), giving as inside surface temperature:

$$\theta_{si} = 0.225\,\theta_e + 0.775\,\theta_i \tag{8.18}$$

i.e. a temperature ratio 0.775. At 21 °C indoors, freezing inside now starts at –72 °C outside. At 0 °C outdoors, the inside surface will touch 16.3 °C. Vapour pressure indoors so must pass 1852 Pa before surface condensate will deposit. Thermal comfort further improves. For a view factor 0.5 with the glass, comfort dissatisfaction now requires a temperature of –50 °C outdoors! But, the mould complaint likelihood in non-insulated or thermal bridge rich building, where low-e double-glass replaces single glass, will become still more pronounced.

Gas filled low-e double glazing

With the radiant exchange in the cavity minimized, slowing down conduction by filling the cavity with a better insulating gas than dry air becomes worthwhile. Most appropriate are the (dry) inert gases argon, krypton, and xenon (Figure 8.7), with thermal conductivities at 10 °C of:

Gas	λ W/(m · K)
Air	0.0250
Argon	0.0168
Krypton	0.0090
Xenon	0.0065

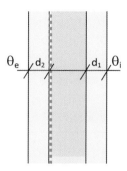

Figure 8.7. Gas filled low-e double glass.

Table 8.8 lists the central thermal transmittance for a cavity width 12 mm and long wave emissivity of one of the bounding surfaces 0.1 or 0. Manufacturers actually guarantee a central thermal transmittance of 1.1 to 1.3 W/(m² · K) for argon and 0.9 to 1.1 W/(m² · K) for krypton.

Table 8.8. Gas filled low-e double glass (6–12–6):
central thermal transmittance for one bounding surface reflective.

Gas	U_{gl}- value W/(m² · K)			Temperature ratio
	e			–
	0.1	0	Actually	
Argon	1.40	1.13	1.1	0.86
Krypton	1.00	0.66	0.9	0.88

For the temperature ratio, see the table. At 21 °C indoors, freezing inside only starts at –126/–154 °C outside. At 0 °C outdoors, the inside surface temperature will reach 18.1/18.5 °C. Vapour pressure indoors so must pass 2071/2128 Pa before surface condensate will deposit. For a view factor of 0.5 with the glass thermal comfort is guaranteed until temperatures outdoors near –80 to –90 °C!

8.2 Glass

Gas filled low-e triple glazing

A central thermal transmittance just below 1 W/(m² · K) is not the end, as heat losses remain higher than across optimally insulated opaque walls, even when counting the solar gains the glass allows. Triple glass with an inert gas fill in the two cavities and one or both bounding surfaces in each low-e was therefore launched. For two 12 mm wide cavities, each with a low-e surface, argon fills assure a central thermal transmittance 0.75 W/(m² · K), while krypton brings the value down to 0.55 W/(m² · K). Two bounding surfaces per cavity low-e gives still lower values. Drawbacks are greater thickness, higher weight and higher edge spacer fracture risk. To limit these disadvantages, a visual light + IR transparent foil could replace the middle glass pane (Figure 8.8). This allows wider convection-free cavities, while compared to double glass the weight hardly increases. For two 20 mm wide cavities with one of them a low-e surface, argon guarantees a central thermal transmittance of 0.62 W/(m² · K), while krypton gives ≈ 0.4 W/(m² · K). Keeping the central foil perfectly stretched however is a challenge.

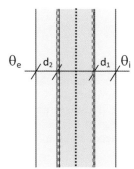

Figure 8.8. Gas filled low-e triple glass with as middle pane a transparent foil.

Additional developments

Edge spacers with lower linear thermal transmittance are slowly gaining acceptance. Some manufacturers further fill double glass with transparent areogel (SiO_2). Vacuum sucking then lowers thermal conductivity to 0.01 W/(m · K). A 15 mm wide cavity then guarantees a central thermal transmittance 0.6 W/(m² · K).

Pure vacuum double glass has been the subject of long-lasting research. As conduction is eliminated and cavity width does not affect radiation, a 0.15 mm wide one with a low-e coating at both surfaces suffices for a central thermal transmittance 0.62 W/(m² · K). However, without separators, the 1 bar overpressure will close the cavity, so, synthetic separators on a 25 × 25 mm² grid have to keep it open. With a 0.15 mm height and 0.4 mm diameter, these create a thermal conductance 0.42 W/(m² · K) in parallel with the cavity's 0.7 W/(m² · K). That way, total thermal transmittance increases to ≈ 0.94 W/(m² · K), for a thickness half the one of double glass. To assure edges tightness, both panes are melted together (Figure 8.9).

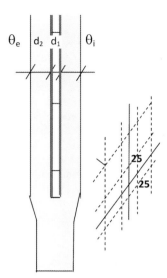

Figure 8.9. Vacuum glass.

Transient response

In transient conditions, neither temperature damping of the glazing nor its dynamic thermal resistance and admittance play any role. Due to the limited thickness and therefore low weight, glazing reacts steady state even at smaller than hourly intervals. However, the solar transmittance, given by the ratio between transmitted solar radiation plus indirect solar heat gain and incident solar radiation has a direct impact on the transient response of spaces and buildings:

$$g = \frac{\tau_K \, E_{ST} + \Phi_{indirect}}{E_{ST}} \tag{8.19}$$

Solar transmittance is highest for a solar beam normal and lowest for a solar beam parallel to the glass. The value for diffuse radiation demands integration over 180°. A constant value equal to the normal incidence one multiplied by 0.95 represents overall solar gains quite well. Table 8.9 lists the property for a few glazing and two solar control glass types.

Table 8.9. Glazings: solar transmittance (g-value).

Glass	G	direct %	indirect %
Single (d = 6 mm)	0.84	98	2
Double (DG)	0.76	89	11
Triple	0.67	82	18
Low-e	0.60	72	28
Low-e, argon filled	0.61	72	28
Low-e, krypton filled	0.61	72	28
Heat absorbing DG[1]	0.28–0.46	79	21
Reflecting DG	0.25–0.40	96	4

[1] absorbing pane outside. If otherwise, direct 54% and indirect 46%

8.2 Glass

A higher value stands for more solar gains and less energy demand for heating. A lower one helps avoiding overheating and economizes on net cooling demand.

Increasing solar absorptivity or reflectivity lowers the solar transmittance of glass. The first demands addition of ferrous oxides, the second evaporation of a thin layer of gold or silver on the glass surface. The solar transmittance of single glass shows that both are effective:

$$g = \underbrace{\tau_K}_{\text{Direct}} + \underbrace{\frac{\alpha_S h_i}{h_i + h_e}}_{\text{Indirect}} \quad (8.20)$$

When short wave absorptivity (α_S) increases with a value x without change in reflectivity (ρ_S), then the direct gain decreases due to the same drop x in transmissivity. Not so for the solar transmissivity. In fact, the higher absorptivity simultaneously lifts the indirect gains with a value 0.26× (the sum $\tau_K + \alpha_K + \rho_K$ stays 1, thus, when $\tau_{S,2}$ equals $\tau_S - x$ and ρ_S stays constant, $\alpha_{S,2}$ must equal $\alpha_S + x$). The result is a 0.74× drop in solar transmittance. When instead reflectivity (ρ_S) drops with a value x without a change in absorptivity (α_S), then solar transmittance effectively decreases by a value x. In other words, making glass reflective is more efficient than increasing absorptivity, see Table 8.9. Indirect gains of course pose less comfort problems as they only increase operative temperature, whereas the direct ones may overheat the bodies they touch.

Both glass types have drawbacks. Absorbing glass heats strongly when radiated. If radiated partially, thermal stresses between warm and cold may reach such high values that rupture follows. Absorbing glass must therefore be hardened. Reflective glass in turn acts as a solar mirror with the reflected solar radiation not only warming the environment but also causing blinding.

Contrary to variable solar shading systems, the actual solar control glass does not allow varying solar transmittance depending on more gains desirable or overheating reprehensible. This may change in the future with the introduction of electro chromic glasses, which have variable solar transmittance.

Solar shading systems can be mounted inside, outside or in between the glass panes (Figure 8.10). Of the three, interior shading is least, exterior shading most effective. This is logical because interior shading first allows the direct and indirect gains to enter before turning them to the outside. As a result, more will be absorbed by the glass while some will warm the shading, both increasing the indirect gains. Exterior shading instead halts most of the radiation before it enters. For the solar transmittances, see Table 8.10.

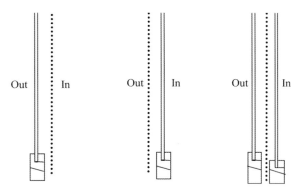

Figure 8.10. Solar shading: inside, outside and in between.

Table 8.10. Solar shading systems: resulting solar transmittance.

Solar shading system		g	Direct %	indirect %
Interior,	reflecting	0.24–0.57	75–96	25–4
	glass fabric	0.57		
Exterior,	roller shutter	0.05–0.10		
	roller blind	0.08–0.27	89–93	11–7
	roller fabric	0.17–0.21		
In between		0.13–0.36		

A simplified calculation for a combination of glass panes and shading starts by writing the thermal balance per layer or pane with the absorbed solar radiation E_{Sa} as a source term. For an exterior roller fabric in combination with double-glazing, the balances look like:

Roller fabric

$$E_{Sa3} + h_e\left(\theta_e - \theta_3\right) + h_{cc2}\left(\theta_{c2} - \theta_3\right) + 5.67\, F_{T2.3}\frac{\theta_2 - \theta_3}{\dfrac{1}{e_2} + \dfrac{1}{e_3} - 1} = 0$$

with E_{Sa3} solar radiation absorbed by the roller fabric.

Cavity between roller fabric and double glass

That cavity has an effect only when air-washed:

$$h_{cc3}\left(\theta_3 - \theta_{c2}\right) + h_{cc2}\left(\theta_2 - \theta_{c2}\right) = c_a\, G_{a2}\frac{d\theta_{c2}}{dt}$$

Exterior glass pane

We assume the pane is isothermal along its thickness and surface:

$$E_{Sa2} + h_{cc2}\left(\theta_{c2} - \theta_2\right) + 5.67\, F_{T2,3}\frac{\theta_3 - \theta_2}{\dfrac{1}{e_2} + \dfrac{1}{e_3} - 1} + \frac{\lambda_{gl}\, X}{d_{gl}}\left(\theta_1 - \theta_2\right) + 5.67\, F_{T1,2}\frac{\theta_1 - \theta_2}{\dfrac{1}{e_1} + \dfrac{1}{e_2} - 1} = 0$$

How much radiation the pane will absorb (E_{Sa2}), depends on the transmissivity of the roller fabric, which is determined by its openness for visual light.

Interior glass pane

Again, the pane is considered isothermal along its thickness and surface:

$$E_{Sa1} + \frac{\lambda_{gl}\, X}{d_{gl}}\left(\theta_2 - \theta_1\right) + h_i\left(\theta_i - \theta_1\right) + 5.67\, F_{T1,2}\frac{\theta_2 - \theta_1}{\dfrac{1}{e_1} + \dfrac{1}{e_2} - 1} = 0$$

with E_{Sa1} the solar radiation absorbed by the pane.

8.2 Glass

Solving the system for $\theta_e = \theta_i = 0$ gives $\Delta\theta_1$, the increase in temperature of the interior pane due to insolation. The solar transmittance then becomes:

$$g = \tau_{S,res} + h_i \frac{\Delta\theta_1}{E_{ST}} \tag{8.21}$$

where $\tau_{S,res}$ represents the resulting short wave transmissivity of the couple glass/shading, which is approximately given by $\tau_{S,res} = \Psi_{vl}\,\tau_{S,1}\,\tau_{S,2}$ with Ψ_{vl} the ratio between perforated and total area of the roller fabric.

Moisture tolerance

Two problems remain: surface condensation and interstitial condensation in multiple glazing.

Surface condensation

As indicated, surface condensation risk decreases the better the glass insulates. Table 8.11 and 8.12 illustrate this for a normally inhabited living room in a social dwelling.

Surface condensation centrally on the inside pane of better insulating glazing systems clearly starts at a relative humidity so high that it no longer warms for mould to form on opaque surfaces, as single glass did.

Table 8.11. Social dwelling, living room, moderate climate: relative humidity, above which surface condensation starts centrally on the inside glass pane.

Glazing	Relative humidity (%)				
	Inside temperature 21 °C, volume 75 m³, $n = 0.5\ h^{-1}$				
	θ_e (°C)				
	−10	−5	0	5	10
Double	49	55	62	70	78
Low-e	65	70	75	*80*	*86*
Low-e, argon	71	75	*80*	*84*	*89*
Low-e, krypton	79	*82*	*85*	*88*	*92*

In italics: relative humidity beyond the threshold for mould on inside partitions

Table 8.12. Social dwelling, living room, moderate climate: average vapour release allowable to avoid surface condensation centrally on the inside glass pane.

Glazing	Vapour release allowable in kg/day				
	Inside temperature 21 °C, volume 75 m³, $n = 0.5\ h^{-1}$				
	θ_e (°C)				
	−10	−5	0	5	10
Double	6.31	6.53	6.53	6.55	6.46
Low-e	8.88	8.86	8.57	8.26	7.73
Low-e, argon	9.89	9.77			
Low-e, krypton	11.09				

Table 8.12 shows that surface condensation centrally on the inside pane requires important vapour releases. Six people using the living room 24 hours a day produce 6.31 kg! Ten people produce 11.1 kg.

The figure for sleeping rooms is less beneficial. As few people heat sleeping rooms in moderate climate regions, average inside temperature stays low. People also close curtains at night with a lower inside surface film coefficient between sleeping room and glass as a result. At the same time in a parent's room, vapour release equals the amount two people produce per hour (≤ 80 g/h). For the consequences, see Table 8.13.

Table 8.13. Social dwelling, parent's sleeping room, moderate climate: average vapour release allowed at night to avoid surface condensation centrally on the inside glass pane.

Glazing, curtains closed	Vapour release allowable at night in g/h				
	Volume 40 m³, $h_i = 4$ W/(m²·K), $n = 0.5$ h⁻¹				
	θ_e (°C)				
	−10	−5	0	5	10
	θ_i (°C)				
	8.6	10.7	12.8	14.9	17.0
Double	*43*	*47*	*46*	*54*	64
Low-e	*59*	*63*	*63*	*70*	77
Low-e, argon	*79*	83	82	87	92
Low-e, krypton	85	89	87	92	96

Italics: vapour release below 80 g/h

What to do? The preference goes to better ventilation, not 0.5 ach but 1 ach, as it doubles the admissible vapour release, while assuring the minimum ventilation needed for indoor air quality reasons. Another, less efficient measure is exterior roller blinds. That way, the outside surface film coefficient drops, which makes the glass warmer. However, an exterior roller blind may keep the room insufficiently ventilated.

With the advent of low-e, gas filled glass a new complaint emerged in moderate climates: condensation outside blurring the view to outdoors. The reason is under cooling during unclouded nights, causing a drop in outside surface temperature under the dew point of the outside air. To give an example, take a vertical low-e, argon filled 5/15/5 mm double glass. Assume the air temperature at night drops to 0 °C. Under windless clear sky conditions, the sol-air temperature will touch −4 °C, dropping the outside surface temperature to −2.1 °C. Surface condensation outside will then deposit at the glass for relative humidity outdoors passing 84%, a value easily touched in moderate but humid climates. The phenomenon disturbs building users. However, they have to accept the situation because nobody can change every day physics and drying the outside air is not an option.

Interstitial condensation

Interstitial condensation in double and multiple glass happens when the edge spacers lose tightness. In case this happens, condensate will accumulate at the coldest cavity bounding surface where it gradually hinders the view through the glass and degrades the building's appearance.

8.2 Glass

Gas-filled low-e glass also sees gas diffusing out and air diffusing in, resulting in a slow central thermal transmittance increase from low-e, gas-filled to the low-e, air filled. Anyway edge spacer rupture marks the end of the multiple glazing's service life.

Thermal bridges

Edge spacer linear thermal transmittance

The central thermal transmittance only represents the multiple glazing's centre area. At the edges, the spacers induce thermal bridging. An upper limit for the linear thermal transmittance (ψ) gives the following reasoning: Take double glass, assume the cavity has an infinite thermal resistance and the spacers an infinite thermal conductivity. Then, conduction along both glass panes (y) yields:

Outer pane:

$$\lambda_{gl} d_1 \frac{d^2\theta_1}{dy^2} + h_i (\theta_i - \theta_1) = 0$$

Inner pane:

$$\lambda_{gl} d_2 \frac{d^2\theta_2}{dy^2} + h_e (\theta_e - \theta_2) = 0$$

Solving both second order differential equations with $\theta = \theta_0$ for $y = 0$ and $\theta = \theta_i$ or θ_e for $y = \infty$ as boundary conditions gives:

$$\theta_1 = \theta_i + (\theta_0 - \theta_i) \exp\left(-\sqrt{\frac{h_i}{\lambda_{gl} d_1}} y\right)$$

$$\theta_2 = \theta_e + (\theta_0 - \theta_e) \exp\left(-\sqrt{\frac{h_e}{\lambda_{gl} d_2}} y\right)$$

At the spacer, heat flow in both panes must be identical though opposite:

$$\Phi_1 = \lambda_{gl} d_1 \left(\frac{d\theta_1}{dy}\right)_{y=0} = -\Phi_2 = \lambda_{gl} d_2 \left(\frac{d\theta_2}{dy}\right)_{y=0}$$

or:

$$(\theta_0 - \theta_i) \sqrt{\lambda_{gl} d_1 h_i} = -(\theta_0 - \theta_e) \sqrt{\lambda_{gl} d_2 h_e}$$

With $a_1 = \sqrt{\lambda_{gl} d_1 h_i}$ and $a_2 = \sqrt{\lambda_{gl} d_2 h_e}$, temperature θ_0 becomes:

$$\theta_0 = \frac{a_1 \theta_i + a_2 \theta_e}{a_1 + a_2} \tag{8.22}$$

which means the upper limit for the linear thermal transmittance is:

$$\psi = \frac{a_1 a_2}{a_1 + a_2} = \frac{\sqrt{\lambda_{gl} d_1 h_i} \sqrt{\lambda_{gl} d_2 h_e}}{\sqrt{\lambda_{gl} d_1 h_i} + \sqrt{\lambda_{gl} d_2 h_e}} \tag{8.23}$$

The upper limit value depends on glass pane thickness and the in- and outside surface film coefficient. For a pane thickness of 5 mm, $h_i = 7.7$ W/(m² · K) and $h_e = 25$ W/(m² · K) we get: $\psi = 0.126$ W/(m · K).

For real multiple glazings, the kind of glass and window type establish the value. It should move toward the upper limit when the glazing system insulates better. Table 8.14 collects data calculated using software for three-dimensional heat flow and lists the values given in the ISO-EN standards. The importance of the effect of the spacers on total thermal transmittance of multiple glazing is shown in Table 8.15.

Applying double and multiple glass in small sizes hardly makes sense, except if spacers with very low linear thermal transmittance could be used, see Figure 8.11.

Table 8.14. Edge spacers: linear thermal transmittance (calculated values in italics).

Glazing	ψ (W/(m · K))			
	Window frame			
	Timber	PVC	Metal without thermal cut	Metal with thermal cut
Double				*0.029*
				0.027
				0.018
	0.06	0.06	0.02	0.06
Tripple	*0.038*			
	0.11	0.11	0.05	0.11
Double, low-e	0.11	0.11	0.05	0.11
Double. Low-e, argon				*0.08*
				0.054
	0.11	0.11	0.05	0.11
Double, low-e, krypton	0.11	0.11	0.05	0.11
Double, low-e, krypton Insulating spacer				*0.04*
	0.07	0.07	0.04	0.07

Table 8.15. Multiple glass: total thermal transmittance as function of size (aluminium frame with thermal cut, standard values for ψ).

Glazing	Size m²				
	1 × 1	0.5 × 0.5	0.25 × 0.25	0.2 × 0.2	0.125 × 0.125
Double	3.04	3.28	3.76	4.0	4.72
Double, low-e	2.34	2.78	3.66	4.1	5.42
Double, low-e, argon	1.54	1.98	2.86	3.3	4.62
Double, low-e, krypton	1.34	1.78	2.66	3.1	4.42

8.2 Glass

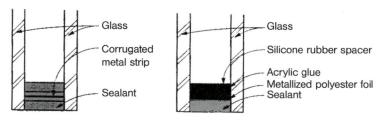

Figure 8.11. Two examples of better insulating spacers (see reference 8.28).

Edge spacer temperature ratio

The lowest value of the temperature ratio follows from:

$$f_{h_i} = \frac{a_1}{a_1 + a_2} \tag{8.24}$$

For double glass with a pane thickness of 5 mm, $h_i = 7.7$ W/(m² · K) and $h_e = 25$ W/(m² · K) the value is 0.36, i.e. hardly above single glass. Therefore better insulating glass systems are not immune to surface condensation. Surface condensation will appear all around the perimeter at much lower relative humidity than calculated in Table 8.11. Of course, Equation (8.24) is too pessimistic. Measurements on glass in an aluminium frame with thermal cut gave higher values, see Table 8.16. Figure 8.12 clearly shows the edge spacer effect.

Table 8.16. Edge spacer: measured lowest temperature ratio.

Glazing	Window frame
	Aluminium with thermal break
Double	0.49
Double, low-e, argon	0.58

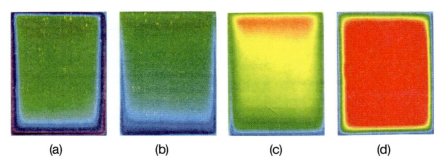

Figure 8.12. Infra-red pictures of different glazing
(a: double with aluminium edge spacers, b: double with better insulating edge spacers,
c: low-e double with better insulating edge spacers, d: super insulating glass) (see reference 8.28).

8.2.2.3 Building physics: acoustics

As façade sound transmission loss requirements, we have:

Environment	Equivalent sound pressure level dB(A)	R_{fac} dB(A)
Rural, sub-urban	$L_{aeq} \leq 55$	No requirement
Urban residential	$55 < L_{aeq} \leq 65$	$22 < R \leq 27$
Light industry, mixed commercial and residential	$65 < L_{aeq} \leq 75$	$27 < R \leq 32.5$
City centres, heavy industry, heavy traffic	$L_{aeq} > 75$	$32.5 < R \leq 37.5$

Glazing usually figures as the weak link. Façade sound transmission loss requirements therefore often reduce to glazing requirements, as the equation for the façade (R_{fac}) proves:

$$R_{fac} = 10 \log \left(\frac{\sum A_i}{\sum \frac{A_i}{10^{0.1 R_i}}} \right) \qquad (8.25)$$

Take a 35 m² large massive façade, $R_{500} = 50$ dB, whose glazing surface equals 1/5 of this area. (8.25) becomes:

$$10 \log \left(\frac{35}{0.00028 + \frac{7}{10^{0.1 R_{glas}}}} \right)$$

Depending on the sound transmission loss of the glass, the façade sound insulation totals:

$R_{glazing}$ dB	R_{fac} dB
20	26.9
25	31.9
30	36.9
35	41.9

From a sound insulation point of view, double glass with panes of equal thickness performs badly. The reasons are mass/spring resonance and coincidence. Both cause a pronounced drop in sound transmission loss, the first at mid-frequency, the second around 3000 Hz, see Figure 8.13. As rule of the thumb, double glass performs hardly better than single glass with the same thickness as one of the panes. A low-e coating doesn't help. Argon and krypton give some relief though very moderately. Increasing internal damping by replacing the dry

8.2 Glass

cavity air by a more viscous gas or layering the glass panes using a synthetic intermediate makes a difference (see Table 8.17 and Figure 8.14). Lowering the resonance frequency also helps, as does increasing the loss left at resonance or separating coincidence of both panes. A lower resonance frequency demands wider cavities (which complicates manufacturing) or thicker glass panes. Panes of different thickness cause a higher loss at resonance and separate coincidence (Figure 8.14).

Combining all this with a low-e coating and a cavity filled with a softer, better insulating gas, allows killing two birds with one stone: a lower central thermal transmittance and a higher sound transmission loss. Still better acoustical performances demand twofold windows.

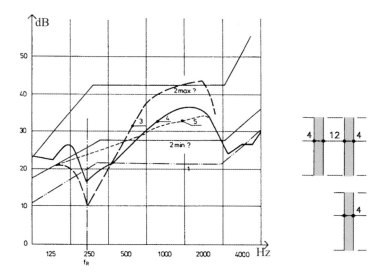

Figure 8.13. Sound transmission loss of glass
(1: single glass, prediction; 2: double glass, maximum and minimum;
3: double glass, impact cavity resonance; 4: double glass, measured; 5: single glass, measured.

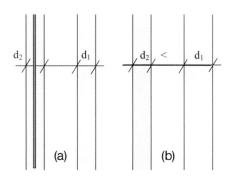

Figure 8.14. Sound insulating double glass:
(a) pane with damping intermediate layer; (b) panes of different thickness.

Table 8.17. Acoustically upgraded double glass: mean sound transmission loss (125–4000 Hz).

Glazing	R_{gl} dB
Double, 4–12–4	27.0
Double, 6–12–4	37.0
Double, 8–12–5	39.0
Double, 6–12–4.5/0.75/4.5	37.0
Double, 12–12–4.5/0.75/4.5	41.0

8.2.2.4 Building physics: light

Insolated clear glass offers the highest illuminance efficiency per m² of all lighting systems: 107 to 149 lux per W incident radiation. Visible transmittance with as symbol LTA (light transmission absolute), given by the ratio between the light transmitted by and the light incident on the glass, both weighted with human eye sensitivity, characterizes the transmitted light:

$$\text{LTA} = \frac{\int_{0.38}^{0.76} \tau(\lambda) E_s(\lambda) O(\lambda) \, d\lambda}{\int_{0.38}^{0.76} E_s(\lambda) O(\lambda) \, d\lambda} \qquad \text{(wavelength } \lambda \text{ in µm)} \qquad (8.26)$$

Because solar radiation contains ±50% visible light, visible transmittance cannot be zero once solar transmissivity (τ_S) of any glazing exceeds 0.5. At a solar transmissivity of 1, visual transmittance must also be 1. Conversely, when visual transmittance exceeds zero, solar transmissivity must do the same, while for a visual transmittance 1, solar transmissivity at least equals 0.5. The basic relationship between the two must therefore look as depicted in Figure 8.15.

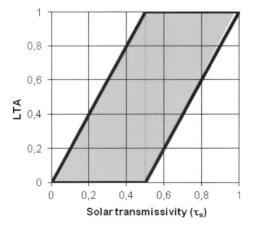

Figure 8.15. Basic relationship between the visual transmittance and solar transmissivity.

8.2 Glass

Table 8.18. Better insulating glazings: visible transmittance.

Glazing	LTA
Double	81
Double, low-e	74
Double, low-e, argon	74
Double, low-e, krypton	74
Double, absorbing	40–65
Double, reflecting	50–55

The art with solar control glass consists of combining acceptable visible transmittance with lower solar transmittance. How well glazing does this, is shown by the ratio between the two: the higher, the better. Actual low-e double glass shows a ratio 1.4, normal double glass a ratio close to 1. Table 8.18 lists a few visible transmittance values.

8.2.2.5 Durability

Manufacturing double and multiple glazing happens at a given temperature (θ_o) and atmospheric pressure (P_{ao}). Once the edge spacers are closed, the cavities act as springs coupling the panes. A change in atmospheric pressure compared to production compresses or expands that spring until the sum of the spring pressure and back pressure exerted by the bending panes equals the outside value. Underpressure compared to production causes tensile stress, while overpressure causes compressive stresses in the spacers. Simultaneously, fluctuating cavity temperatures add gas pressure fluctuations with additional pane bending and tensile or compressive stresses in the spacers (Figure 8.16).

Besides, temperature difference between the glass panes induces distinct expansion and contraction of each pane, which in turn shears the spacers (Figure 8.16). Loads and related spacer stresses are dynamic in nature. Fatigue therefore causes fracture, which is classified as an aging phenomenon. But it surely makes service life of double and multiple glazing function of slope, orientation and the care taken during production and mounting. Installing multiple glazing too stiffly and narrowly excludes movement and increases future stresses in panes and spacers.

Whereas the air pressure effect does not depend on the central thermal transmittance, temperature effect does as a lower value increases temperature difference between panes. That enlarges expansion and shrinkage and lifts shear in the spacers. Also cavity temperatures differ somewhat. Tables 8.19 and 8.20 show the two effects. Particularly in winter a lower central

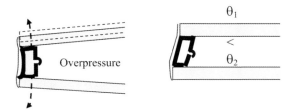

Figure 8.16. Double glass: tension, compression, bending and shear at the edge spacers due to changing cavity temperatures, different pane temperatures and changes in atmospheric pressure.

thermal transmittance scores negatively. Compared to normal double glazing, temperature difference between the two panes increases 1.6 times for low-e, krypton filled double glazing., while underpressure in the cavity touches 4000 Pa, yes, even 7600 Pa during a cold winter day. To compare with, high wind speeds are needed to see 600 Pa windward pressure. Of course, internal pressure changes to the extend the panes bend. A volume reduction of the cavity with 8% suffices to cut a 2600 Pa underpressure. Related glazing deformation anyway induces compound bending in the spacers.

Table 8.19. Double glazing: mean temperature in panes and cavity during a cloudy winter day ($\theta_i = 21$ °C, $\theta_e = 0$ °C, produced at 20 °C, 1 bar).

Double glazing	Temperature (°C)			Shrinkage (μm/m)		Tension (MPa)		Under-pressure
	Glass		Cavity	(free moving)		(glass restrained)		Cavity Pa
	Pane 1	Pane 2		Pane 1	Pane 2	Pane 1	Pane 2	
Normal	2.7	13.3	8.0	150	50	10.9	4.2	4080
Low-e	1.7	16.3	9.0	160	30	11.5	2.4	3760
Low-e, argon	1.3	17.3	9.3	160	20	11.8	1.7	3970
Low-e, krypton	0.9	18.4	9.6	170	10	12.0	1.0	3860

Table 8.20. Double glazing: maximum and minimum temperature in the panes during a summy cold winter day and sunny hot summer day.

Double glazing		Glass temperature			
		Cold winter day		Hot summer day	
		Pane 1, °C	Pane 2, °C	Pane 1, °C	Pane 2, °C
Normal	Minimum	−12.6	7.3	17.9	22.3
	Maximum	−6.3	13.5	30.2	30.4
Low-e	Minimum	−14.4	12.6	17.4	23.9
	Maximum	−8.1	18.7	30.2	30.5
Low-e, argon	Minimum	−15.0	14.3	17.3	24.3
	Maximum	−8.7	20.5	30.2	30.5
Low-e, krypton	Minimum	−15.7	16.3	17.1	24.8
	Maximum	−9.4	22.5	30.2	30.6

8.2.2.6 Fire safety

Glass is an elastic material that absorbs little strain. Fire increases temperatures so quickly that it ends in brittle rupture, followed by flame flashover. Retarding demands glass with low thermal expansion coefficient (boron silicate glass, $\alpha = 3 \cdot 10^{-6}$ m/m instead of $9 \cdot 10^{-6}$ m/m for normal glass, glass ceramic with $\alpha = 0.1 \cdot 10^{-6}$ m/m), pre-stressing the glass or applying layered or reinforced glass (Figure 8.17).

8.2 Glass

Figure 8.17. Reinforced glass.

If the glazing should stay flame proof and slow-down heat transfer, double glass, in which a transparent, protecting material such as sodium silicate fills the cavity, can be used. If the pane at the fireside breaks, the sodium silicate foams and forms a non-burnable insulation layer.

8.2.2.7 Break-in safety

Again good deformability matters. This demands layered glass, a measure making it even bulletproof. With double-glass, the layered pane is located at the impact side, for glass in envelopes the outer side. Glazings are also more often integrated in the overall safety system, for example by installing surface contact sensors.

8.2.3 Technology

Performance analysis showed that low-e, gas filled double, and multiple glass can meet the requirements imposed. The question of how to deposit the low-e coating and fill the cavity with other gasses has actually been solved. One problem left is the edge spacers. These have to stay gas tight, mechanically strong, deformable, thermally insulating, sound damping, etc. The first double glass generation had lead spacers, adhered to the panes with copper. Service life was limited, on the average some 10 years, though many were in service for 30 years and more (Figure 8.18a). Later, aluminium profiles, filled with silica gel and adhered to the panes using a synthetic intermediate such as butylene, polyurethane, or silicon rubber, took over (Figure 8.18b). Of these three intermediates, silicon is the least gas tight. Of course, the rubbers upgraded spacer deformability and prolonged service life in terms of air tightness compared to the lead spacers. A thermal bridging effect however remained. This became worse as the glazing gained in insulation quality (see above). With the advent of low-e gas filled double glass, the search for spacers with better thermal properties started. Promising are spacers of silicon rubber with aluminium foil as a gas tight layer.

Another research subject is glass with varying solar transmittance: high at limited, low at abundant insolation. Also self-cleaning glass is being examined. Sometimes a meaningful development needs to find the right application. An example is heated glass: excellent for premature incubators in hospitals but not for home heating. Using the inner pane as heating surface eliminates the surface film resistance inside as additional insulation, while the pane keeps temperatures quite above the indoor comfort value. Even the best glazing types lose in

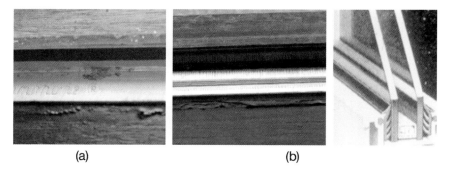

Figure 8.18. Edge spacers:
a = lead with copper bond, b = aluminium with neoprene bond and silica gel fill.

terms of thermal resistance in comparison to well-insulated opaque assemblies. Higher inner pane temperatures also reinforce radiation and may initiate convection in addition to more conduction in the glass cavity. The result is more energy used than with other heating systems. Moreover, electrical heating cannot be recommended in terms of primary energy consumed for countries lacking hydraulic, solar or wind turbine electricity.

8.3 Windows and doors

8.3.1 In general

Until the middle ages, building openings were filled with shutters instead of windows. Later came timber framed leaded glass, providing protection against rain and wind while allowing some contact with outdoors. The range of window types today is very broad. They combine fixed and operable sashes, allowing peak ventilation. Depending on how they project or slide, we distinguish the following types (Figure 8.19):

- **Casement windows.** The operable sashes have hinges at one of the vertical sides and open inwards or outwards around that side.
- **Reversible windows.** The operable sash turns halfway around a horizontal axis
- **Fan windows.** The operable sash turns around a horizontal axis, located above or underneath
- **Fan casement windows.** The operable sash acts as fan and casement window
- **Horizontal sliders.** The operable sash slides horizontally in front of or behind the fixed casement
- **Hung slider.** The operable sash slides vertically in front of or behind the fixed casement

The window frame has to prevent walls which contain windows from loading the glass. Windows of course also serve aesthetics. Designers love playing with glass distribution and frame forms. Doors are another subject. When they contain glass, they resemble windows. For non-glazed doors, see Chapter 10.

We expect windows and doors to be lightweight. Otherwise, mounting and manipulating becomes heavy. Wind bends the frame. Withstanding related stress and strain demands a tensile resistive material and correct section modulus. The three requirements – lightweight,

8.3 Windows and doors

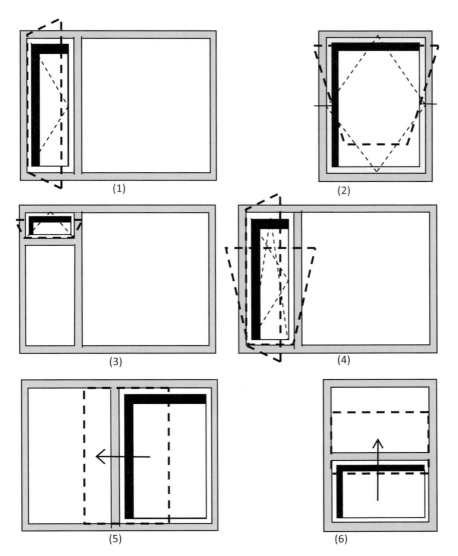

Figure 8.19. Window types: (1) casement window, (2) reversible window, (3) fan window, (4) fan casement window, (5) horizontal slider, (6) hung slider.

tensile resistive, correct section modulus – limit the choice of suitable materials to timber, steel, aluminium and synthetics. The advantages and drawbacks of the three most used, timber, aluminium, and vinyl, are:

- *Timber* (Figure 8.20a)
  Light and easy to process. Has a low thermal conductivity, a low thermal expansion coefficient, and a favourable strength to stiffness ratio (modulus of elasticity 10 000 MPa). Is moisture sensitive. Rot as well as anisotropic deformation may be a problem. Ultraviolet radiation greys timber. Moisture sensitivity and discoloration requires painting or staining.

(a)   (b)   (c)

Figure 8.20. Window frames: a = timber, b = aluminium, c = vinyl.

- *Aluminium* (Figure 8.20b)
  Light, strong and quite stiff (modulus of elasticity 70 000 MPa). Has a very high thermal conductivity and a two times larger thermal expansion coefficient than steel (λ = 230 W/(m·K), $\alpha = 24 \cdot 10^{-6}$ K^{-1}). Compared to timber it is more difficult to process. Window frames are therefore industrially manufactured, starting from extruded sections in anodised aluminium.
- *Vinyl* (Figure 8.20c)
  Softens above 82 °C, is moderately stiff (modulus of elasticity 2500 MPa), has good impact strength, and is, if stabilised for UV, highly weather resistant. It is also easy to extrude and weld, with the welds remaining tough. Drawbacks are the very high thermal expansion coefficient ($70 \cdot 10^{-6}$ K^{-1})) and the degrading stiffness at higher temperatures. The last obliges manufacturers to reinforce vinyl frames with steel sections.

Window and sash frames have rabbets all around. That way one can mount and fix any glazing without fracturing. The width and depth of the rabbet depends on the glass type, its thickness and the fit needed between glass and frame for sealing with putty or a preformed rubber profiles. Setting blocks in the lower rabbet corners helps mounting the glazing, whose dimensions equal the distance between the rabbet downsides minus two times the necessary perimeter fit. These blocks are high as the perimeter fit. The glass is fixed with glazing beads that sit preferentially at the inside (safer against break-in, easier for future glass replacement). Between glass and frame there is a two-steps joint.

Mounting double-glass in a timber window for example happens as shown in Figure 8.21. First, alongside the whole perimeter, a one side adhering foam rubber strip is fixed as spacer at both sides of the glazing. Next, the glazing is put on setting blocks, which sit in the lower rabbet at some 10 cm from the corner. Then the glazing beads first get a putty strip at their underside before screwing them on the window frame. After, a glass sealant of the classes 25LM or 25HM closes the joints between glazing and beads and glazing and frame.

With aluminium and vinyl window frames and sashes, the glass beads are fixed by snapping, while instead of closing the joints between glazing and beads and glazing and frame with a sealant, a preformed rubber profile inside and outside makes the finished rabbets rain, wind and airtight

Door and window hardware include all attributes needed to open and close, i.e. hinges, handles, locks, lock bars, etc., Figure 8.22.

8.3 Windows and doors

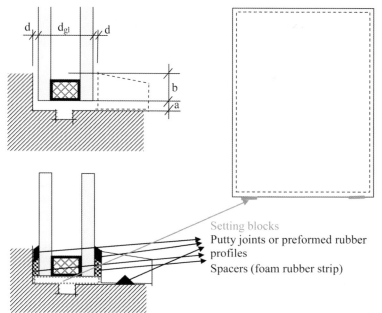

Fits a and d, height b:

Pane thickness	4 mm	5 mm	≥ 6 mm	
Edge spacer	Without exterior stainless steel cover			With
A	≥ 4 mm	≥ 4 mm	≥ 6 mm	≥ 6 mm
D	≥ 3 mm	≥ 3 mm	≥ 3 mm	≥ 3 mm
B	≥ 14 mm	≥ 14 mm	≥ 14 mm	≥ 16 mm

Figure 8.21. Timber window: mounting double glass in the rabbets.

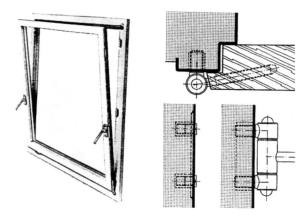

Figure 8.22. Door and window hardware (hinges, handles, etc.).

8.3.2 Performance evaluation

The performance requirements cover everything, from structural safety to building physics, durability, fire safety and maintenance.

8.3.2.1 Structural safety

Glazing and wind are the main loads. Not only should the frame have sufficient resistance, bending of horizontal and vertical sections under extreme wind must also stay below $1/300^{th}$ of the span (control by calculation). For operable parts, the force needed to operate should stay below 100 N, while deformation by shear and forces normal to the window, once open, may not aggravate closing (control by measuring), and the hardware must be fitted for the purpose. The upper hinges of a casement window for example must withstand tension, the lower hinges must withstand compression exerted by the open sash. Handles should be able to withstand repeated usage.

Windows are mounted in a way the opaque enclosure cannot transmit forces and loads to the frames. Therefore, joints, already needed for mounting, separate the two. The frames are fixed to the walls at their lower and side edges, never at their upper edge. If fixed there, coupling must allow shear. Fixations must be strong and stiff enough to transmit the load on the window to the opaque enclosure parts all around.

8.3.2.2 Building physics: heat, air, moisture

Air tightness

Air-tightness is a key requirement for windows and outer doors. Defective air-tightness jeopardizes rain tightness, degrades sound insulation, may give draft complaints, and could induce surface condensation on frames made of hollow sections. Not only are the rebates between fixed parts and sashes critical, but also joints between rails and jambs and, the joints around windows, which should guarantee continuity with the airtight layer in the surrounding enclosure, figure as potential leaks. Windows actually require a certificate in which the manufacturer guarantees the air permeance coefficient per meter run of rebate (a, units $m^3/(m \cdot h \cdot Pa^{2/3})$):

$$a = \frac{\dot{V}_a}{\Delta P_a^{2/3}} \qquad (8.27)$$

with $\dot{V}$ the airflow in m^3 per hour and meter run of rebate and ΔP_a the air pressure difference across the window. The requirements differ between countries, see Table 8.21.

Table 8.22 and Figure 8.23 summarize air permeance coefficients as measured. Both show that only windows with weather-strips in the rebates fulfil the requirements. Moreover, even then, proper fitting and correct mounting is crucial.

8.3 Windows and doors

Table 8.21. Air permeance coefficient per meter run of rebate in $m^3/(m \cdot h)$.

Country		\multicolumn{4}{c}{Class}			
		A^3	B^3	C^3	D^4
Belgium	ΔP_a (Pa)	150	300	500	> 500
	a-value	0.28	0.14	0.096	Specifications
The Netherlands	ΔP_a (Pa)	75	150	300	$= B^1$
	a-value	0.50	0.32	0.20	
	ΔP_a (Pa)	300	300	450	$= K^2$
	a-value	0.20	0.20	0.153	
Germany	ΔP_a (Pa)	150^3	300^3	600^3	> 600
	a-value	0.43	0.215	0.215	Specifications
Switzerland	ΔP_a (Pa)	150^3	300^3	600^3	
	a-value	0.20	0.20	0.20	

[1] B for inland locations
[2] K for coastal locations
(North Holland, West Frisian islands, Ijssel lake, up to 2.5 km from the North sea)
[3] Linked to following building heights: 150 Pa: ≤ 8 m; 300 Pa: $8 < h \leq 20$ m; 600 Pa: $20 < h \leq 100$ m
[4] $h > 100$ m

Table 8.22. Timber and metal windows: measured air permeance coefficients.

Weather-strips in the rebates?	\multicolumn{4}{c}{a in $m^3/(m \cdot h)$ at $\Delta P_a = 1$ Pa}			
	Mean	Stadev	5%	95%
Yes	0.61	0.74	0	2.50
No	2.74	2.02	0.6	7.55

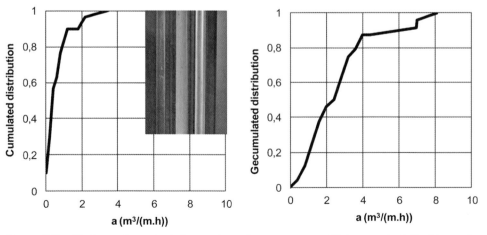

Figure 8.23. Timber and metal windows, rebate air permeance coefficient, on the left with weather-stripping, on the right without.

Thermal transmittance

We introduced the concept 'thermal transmittance of a window' when discussing the glazing. Calculation requires FEM or CVM. If this is impossible, for example because not all window details are given, the values of Table 8.23 can be used.

Compared to timber, vinyl and PUR, aluminium frames perform much worse. Only a thermal break, which consists of a tension and compression resisting material with low thermal conductivity, guarantees acceptable performance. For that to be the case, the distance between the two frame halves must be as large as possible, while the contact between break and aluminium should remain minimal. Much used as materials are fibre reinforced polyamide, $\lambda = 0.4$ W/(m·K), and cast PUR, $\lambda = 0.18$ W/(m·K). Polyamide is applied lamella-wise, PUR block-wise (see Figure 8.24). The wider the break, the lower the thermal transmittance of the aluminium frame. Filling the air space between the polyamide lamellas with PUR-foam yields additional gains. Glass and thermal break should form one plane.

Table 8.23. Window frames: thermal transmittance.

Material	Frame details	U_{fr}, W/(m²·K)
Hard wood	Depends on the thickness d in m of the frame	$\dfrac{0.32}{d + 0.086}$
Soft wood	Depends on the thickness d in m of the frame	$\dfrac{0.25}{d + 0.074}$
PUR		2.8
Vinyl	Three chambers, metal stiffeners	2.0
	Two chambers, metal stiffeners	2.2
Aluminium	Without thermal break	$\dfrac{1}{\dfrac{A_{fr,i}}{h_i \, A_{fr}} + R_{fr} + \dfrac{A_{fr,e}}{h_e \, A_{fr}}}$ with: $R_{fr} = \dfrac{1}{5.9} - 0.17$
	With thermal break, see Figure 8.24 d: smallest distance between the two frame halves (m) $A_{fr,i}$: total surface indoors of the inside frame halve (m²) $A_{fr,e}$: total surface outdoors of the outside frame halve (m²) A_{fr}: projection on a plane parallel to the window, frame surface (m²).	$\dfrac{1}{\dfrac{A_{fr,i}}{h_i \, A_{fr}} + R_{fr} + \dfrac{A_{fr,e}}{h_e \, A_{fr}}}$ with: $R_{fr} = \dfrac{1}{\dfrac{0.0045}{d} - 18\,d + 2.7} - 0.17$

8.3 Windows and doors

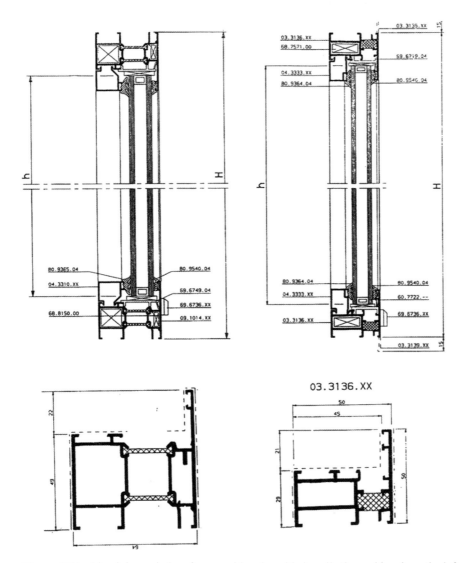

Figure 8.24. Aluminium window frames with polyamide lamella thermal break on the left and with PUR block thermal break on the right.

Outer doors require the same maximal thermal transmittance values as windows, see Table 8.3. If transparent, low-e, gas-filled glass must be used. Otherwise, the door could have a timber or metal frame, the latter with a thermal break and the leaf filled with a thermal insulation material such as mineral wool.

Moisture tolerance

Windows and outer doors face three challenges: rain tightness, surface condensation, in some cases interstitial condensation.

Rain tightness

Rain and air tightness are closely linked. The air permeance categories of Table 8.21 also apply for rain. For windows and outer doors to belong to a given category, no rain penetration may be noted during a standardized rain test at the wind pressure difference characteristic for that category. A two-step rebate between fixed parts and sashes with the outer step as rain control, the inner step as air tightness control and in between a pressure-equalizing chamber is the best way to assure rain tightness. Figure 8.25 show how to realize this when designing a timber window with a sash pivoting to the inside. One starts from two standard timber beam sections, one for the fixed part and one for the sash. Both can slip somewhat against each other in the horizontal (x) and vertical direction (y) with the final position defining the window's outlook. Slipping must observe a few rules. The overlap in y-direction should be at least 3 cm. If less, it excludes correct detailing of the two-step rebate. The slip in x-direction must allow correct design of the pressure equalizing chamber and leave enough distance between chamber and inner rebate, while all contact surfaces between fixed and sash should slope a little to the inside. The pressure-equalizing chamber gets two or more drains to the outside. With a fixed frame and sash in the same plane, an aluminium profile screwed on the fixed part often forms the outer rebate. Vinyl and aluminium frames follow the same principles, a two-step rebate with pressure equalizing chamber in between, see Figure 8.20 above.

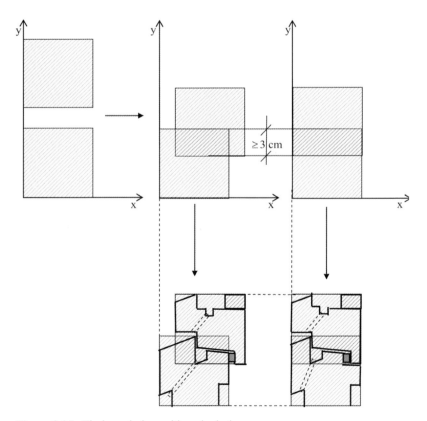

Figure 8.25. Timber window with sash, design.

8.3 Windows and doors

Surface condensation

Surface condensation is a problem with aluminium frames without thermal break. Their thermal transmittance reaches 6 W/(m² · K), somewhat higher than single glass, resulting in the same surface condensation risk as single glass, and even higher with anodised aluminium. Its lower long wave emissivity in fact reduces the inside surface film coefficient and through that the inside surface temperature compared to single glass:

$$\theta_{si} \approx 0.2\,\theta_i + 0.8\,\theta_e \qquad (8.28)$$

The consequence is faster surface condensation on the frame and the risk that plastered reveals will suck the deposit and become prone to mould. The many complaints about this rather than the higher thermal transmittance accelerated the introduction of frames with thermal break, which increased inside surface temperature to:

$$\theta_{si} \approx 0.57\,\theta_i + 0.43\,\theta_e \qquad (8.29)$$

i.e. more or less the same as near to the edge spacers of low-e, gas filled double glass. However, still, be cautions. If for one or another reason both the inside and outside frame halves are in contact with the same high thermal conductivity material, the break short-circuits and the inside halve will be cooler in winter than the equation (8.29) indicates. Also leaky jointing between rails and jambs may allow inside surface condensation. The penetrating cold outside air in fact drops the inside surface temperature around the jointing to $0.4\,\theta_i + 0.6\,\theta_e$ with as a consequence, winter surface condensation despite the thermal break. An efficient remedy manufacturer is closing all jointing with internal corner elements.

Interstitial condensation

Vinyl and aluminium frames, the latter with a thermal break, may suffer from interstitial condensation. The break in fact enlarges temperature difference between both frame halves. In case the frame as a whole lacks air and vapour tightness, for example due to leaky jointings, condensation deposit against the colder half turns real. If the frame allows drainage to the outside, this deposit is hardly harmful. If the frame does not allow drainage, the deposit may cause harm.

Frames combining aluminium outside with timber inside face analogous problems, in this case that of winter condensation of hygroscopic moisture from the timber against the aluminium (Figure 8.26). Also insulated outer doors demand attention. Door leaf assemblies, which are more vapour retarding inside than outside, minimise risk.

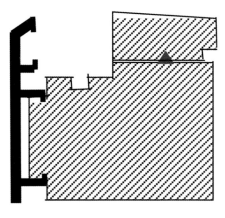

Figure 8.26. Window frame combining aluminium outside with timber inside.

8.3.2.3 Building physics: acoustics

We explained above how to upgrade the sound insulation of glazing. To facilitate good performance, frames must be highly airtight and stiff. Further-on, filling the rabbets with dampening material adds benefit (Figure 8.27a), whereas the joints between the frames and the opaque enclosure should be sealed in two steps, with an elastic sealant in- and outdoors and sound absorbing mineral wool in between (Figure 8.27b). If severe requirements prevail, dual frame and twofold windows are the solutions left. Their noise transmission loss passes 40 dB at 500 Herz, see Table 8.24 and the Figures 8.28.

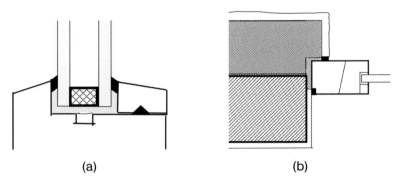

(a) (b)

Figure 8.27. Upgrading sound insulation: (a) rabbet filled with dampening material, (b) two-step joint between frame and opaque enclosure assembly.

Table 8.24. Dual frame and twofold windows: noise transmission loss.

Performance	Dual frame windows	Twofold windows
$35 \leq R_{500} < 40$ dB	Airtight, distance between both glazings ≥ 40 mm	Single or double glass
$40 \leq R_{500} < 45$ dB	Airtight, distance between both glazings ≥ 60 mm, thick glass (combination of double thermally better outside and single inside)	Airtight, single or double glass
$45 \leq R_{500} < 50$ dB	Airtight, distance between both glazings ≥ 100 mm, thick glass (combination of double or thermally better outside and single inside)	Airtight, distance between both windows large, thick glass (combination of double thermally better outside and single inside)
$R_{500} \geq 50$ dB	–	Extremely airtight, distance between both windows as large as achievable, thick glazings (combination of double thermally better outside and single inside), side surfaces between both windows sound absorbing

8.3 Windows and doors

Figure 8.28. Left: Twofold window. Middle and right: double windows.

8.3.2.4 Durability

The three main window frame materials – timber, aluminium, and vinyl – behave differently with changes in temperature and relative humidity. Timber hardly reacts under temperature load but is sensitive to differences and changes in relative humidity. Vinyl instead reacts strongly under temperature load while remaining unaffected by differences and changes in relative humidity. Aluminium only reacts under temperatures load.

These differences have consequences. The way of manufacturing the composite aluminium/timber frame, shown in Figure 8.26, must neutralize all drawbacks of the high thermal conductivity of aluminium. The difference in reaction requires coupling both materials in a way movement remains possible.

Colouring vinyl windows dark is altogether wrong. Insolation in such case warms up the frames much more than it does with white ones, ending in more expansion, more softening and more warping of the vinyl. Avoidance is sought in reinforcing the frames with metal sections.

Due to differences in hygric loading, timber outer doors balloon a little in winter and inflate somewhat in summer. The phenomenon is more pronounced the better the thermal insulation of the door leaf. A solution consists of replacing wooden door boarding by a groove and tongue lathed boarding. Of course, an air and vapour retarder is then needed at the inside of the thermal insulation infill.

8.3.2.5 Fire safety

Windows figure as the weak links here. Façades of mid- and high-rises therefore must be designed so that the length between window rows at floors above one another is at least 1 meter. If windows act as escape routes, pivoting sashes have to leave openings beyond 90 × 90 cm.

8.3.2.6 Maintenance and safety at use

Cleaning windows must be safe. For that reason, high-rises have façade lifts with vertical track guides along the façade. Windows of mid- and low rises are designed in a way that with open sashes the outside of all fixed parts are within reach. For safety reasons, make sashes that pivot to the outside small enough that children cannot wriggle through them. Also handling a window may not turn closing into an equilibriuim act.

8.3.3 Technology

8.3.3.1 Timber, vinyl

The technological upgrade of window frames as a consequence of ever more severe performance requirements started in the 1970s, when better air-tightness and the two-step rebate with weather-strip were introduced. It is anyhow still important to check that all rail and jamb rebates between fixed frame and sashes stay in the same plane, that a drip nose at the sash protects the exterior rebate of the lower rail and that weather-strips are sealed at all corners.

8.3.3.2 Aluminium, steel

Air-tightness and two-step rebates are also important here. A further upgrade of the thermal break remains a challenge. Requirements are: strong and stiff, while still allowing movement of the outer frame half with respect to the inside one, good thermal insulation, and the best durability possible. Possible upgrades are wider breaks and filling the space between the two frame halves with insulating foam.

8.3.3.3 Window blinds and roller shutters

Window blinds and roller shutters were not considered in the performance evaluation. Some elements (Figure 8.29):

- Roller shutters consist of hinged vinyl, timber or aluminium lamellas
- Shutter boxes are mounted build-up or build-in
- With build-in boxes, guide bars are part of the window frames, with build-up boxes, they sit in front of the frame
- Build-in shutter boxes demand thermal insulation at the bottom, the back, and the top, where it has to line up with the outer wall insulation.

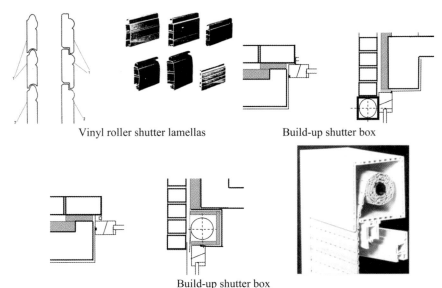

Figure 8.29. Roller shutters.

8.3 Windows and doors

Blinds and roller shutters upgrade the time-averaged thermal transmittance of the window:

$$U_{tot} = \frac{1}{T}\left[U_w(T-\Delta T) + \frac{\Delta T}{\frac{1}{U_w}+\Delta R}\right] \qquad (8.30)$$

with ΔT the period in hours a day they are closed. The other parameter in the formula is the additional thermal resistance offered (ΔR). Defining are air-tightness with the blinds or roller shutters closed and the material these are made from, see Table 8.25. The notions 'very air permeable', 'moderately air permeable' and 'hardly air permeable' in the table relate to the joint width below (b_1), above (b_2) and aside (b_3) the blinds and roller shutters. Of the joints aside, only the widest counts:

	$b_{sh} = b_1 + b_2 + b_3$ mm
Very air permeable	$15 \leq b_{sh}$
Moderately air permeable	$8 \leq b_{sh} \leq 15$
Hardly air permeable	$b_{sh} \leq 8$

For more detailed information we refer to the references and literature.

Table 8.25. Blinds and roller shutters: additional thermal resistance.

Blind or roller shutter	Thermal resistance of the blind or roller shutter $m^2 \cdot K/W$	Additional thermal resistance ΔR $m^2 \cdot K/W$		
		Very air permeable	Moderately air permeable	Hardly air permeable
Aluminium roller shutter	0.01	0.09	0.12	0.15
Timber or vinyl roller shutter, lamellas without foam fill	0.10	0.12	0.16	0.22
Timber or vinyl roller shutter, lamellas with foam fill	0.15	0.13	0.19	0.26
Timber blinds with thickness 25 to 30 mm	0.20	0.14	0.22	0.30

8.3.3.4 Trickle vents

The move to airtight windows harmed adventions ventilation by infiltration, resulting in sometimes nasty mould problems in dwellings. To avoid that problem, national standards specified airflows and the use of purpose designed ventilation systems in residential buildings. Four systems are allowed: natural ventilation, supply ventilation, extract ventilation and balanced ventilation, the last with or without heat recovery. Natural and extract ventilation demand trickle vents above the windows of all day-time and night-time rooms (Figure 8.30).

Figure 8.30. Different trickle vent types.

If these miss a thermal break, thermal transmittance is that of single glass. If present, the vents equal double glass. Other performance requirements are rain-tight, safety against break-in, insect proof, air pressure responding, and easy maintenance. Manufacturers offer a wide variety of trickle vents: with or without thermal break, sound absorbing, hardly visible, as Figure 8.30 on the right shows.

8.4 Glass façades

8.4.1 In general

Complete façades are designed as window systems with studs and transverse beams replacing the jambs and rails. We speak in such cases of glass façades. One infill per floor is called a window front. A curtain wall is a glass façade that covers the whole building height. These include the use of suspended glass. If the glass façade is doubled and functions as a heat exchanger, it is called a double-skin façade. When such a façade has the outside skin covered with PV-cells, it becomes a PV-façade. For the performances of the glass and frame types used, we refer to the sections on glass and windows. The evolutions listed there also hold for glass façades. Only the technology differs.

8.4.2 Window fronts

A window front stands for a floor-high window, as wide as the façade module, with the bays between jambs and rails filled with glass or opaque panels (Figure 8.31).

The last should have thermal resistances comparable to opaque assemblies: 2.5 $m^2 \cdot K/W$ and better. The use of vacuum insulation allows limited panel thicknesses. The average thermal transmittance obeys the expression given for windows, see Equation (8.6). The differences among window fronts concern the surface ratio between glass and opaque panels and the way panel infill is done. The number of possible variants is huge, though some basic principles can be identified:

8.4 Glass façades

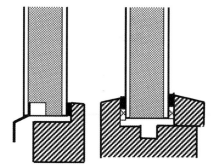

Figure 8.31. Window front, panel mounting.

- Sandwich panels with a vapour-tight finish outside (steel, glass, aluminium) require a vapour tight inside finish from indoor climate classes 2 up to indoor climate class 5.
- Mounting the panels resembles the way done for double-glazing: using setting blocks and applying two-step sealing. Of course this still allows many variants (Figure 8.31).
- Better noise transmission loss follows from using sound-absorbing insulation materials, applying inside and outside finishes with different stiffness and using deformable edge spacers.

8.4.3 Curtain walls

We differentiate between element and stud/transverse beam curtain walls, (Figure 8.32). An element type consists of ready-made, modular panels that, once mounted, form the curtain wall. Each panel is hung separately at the load bearing structure. Once everything is in place, the joints in between are finished. Instead, for a stud/transverse beam curtain wall, mounting starts by fixing the studs at the load bearing structure in a way thermal expansion remains possible and all studs safely transfer the curtain wall's own weight and wind load to the structure. In between the studs come the transverse beams, after which glazing and opaque panels fill the bays formed that way.

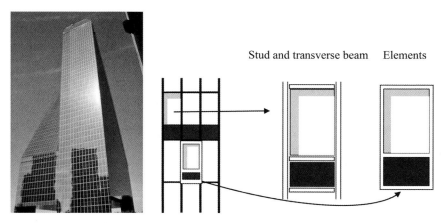

Figure 8.32. Curtain walls.

Structurally, element and stud/transverse beam types behave differently. Element façades transfer their own weight and wind load per panel to the load bearing structure, whereas stud/transverse beam walls first transfer the load to the studs, which transmit it point wise to the load bearing structure. This demands a stud sections with a high section modulus. Box sections fulfil this criterion. Also the transverse beams get a box section, be it with lower section modulus. Together with the cover plates, they deliver the rebates needed to mount glazing and opaque panels (Figure 8.33).

As said, fixing must be such, that thermal movement, including elongation as well as shortening and warping, remains possible. Therefore, one limits the length per stud to a couple of floors, while each stud gets an articulated fixing, sliding and rotating bearings with the building structure.

Air tightness demands a fixed air retarder plane. The problems this causes at the joints between successive stud lengths and between studs and transverse beams requires a technologically complex solution. The use of sliding couplers is an example.

Calculating the thermal transmittance requires software for three dimensional heat transfer. The effect of thermal bridging could be large, which is why the box profiles have a thermal break in the plane formed by the glazing and the opaque infill panels.

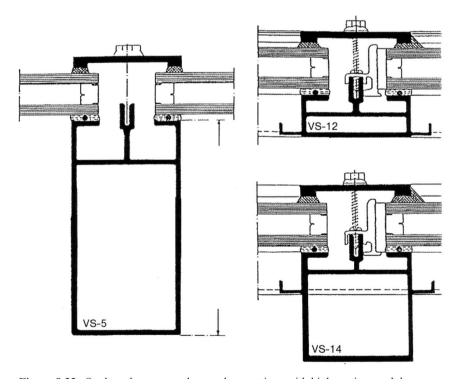

Figure 8.33. Studs and transverse beams: box sections with high section modulus.

8.4.4 Structural and suspended glazing

Quite a recent evolution is the application of structural and suspended glazing. With structural glazing, the glass panes are glued with silicone seals at the studs and transverse beams, see Figure 8.34.

The panes must at least be 6 mm thick. Gluing of course causes problems. For example the glass and the thermal break no longer form one plane. Also from a rain tightness point of view, the choice is delicate as the silicon seals function as one-step joints, turning well-controlled rain drainage into an important asset.

Suspended glazing demands clipped connections to hang the panes at load bearing transverse beams. Because glass has a high tensile strength and buckling does not occur when suspended, the method allows realizing large glazed surfaces crossing several floors. Especially for prestigious buildings, this is an attractive variant for realizing true transparency. However, as with structural glazing, suspended glazing and its stiffeners demand a thorough performance analysis before deciding in favour of its application.

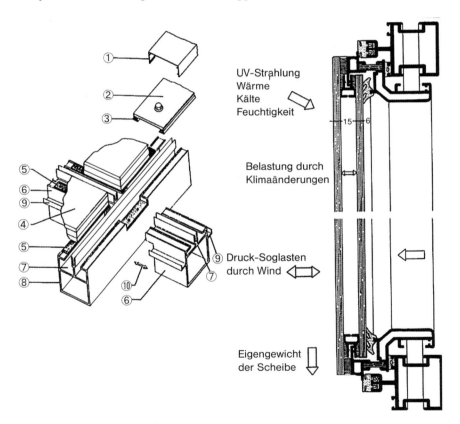

Figure 8.34. Structural glazing.

8.4.5 Double skin façades

8.4.5.1 In general

As we said, double skin façades, sometimes called active façades, function as heat exchangers. In theory, such exchange happens when air infiltrates across air permeable opaque assemblies. The alternative is air flowing longitudinally from inside to outside or vice versa between two leafs forming a double façade. That is the way a double skin façade, consisting of two glass sections, separated by a wide air space forming the flow path, functions. The typology, summarized in Table 8.26, is based on façade lay-out, what air passes through, if the flow is stack induced or forced and if acting as air inlet, air outlet or only as a curtain flow of inside or outside air. The last means inside air is extracted by a central air handling unit along the air space or the air space is flushed by outside air.

Table 8.26. Double skin façades: a typology.

Typology	Which air?		Flow?		Function	
	Outdoor	Indoor	Stack induced	Forced	Supply, extract	Air curtain
Continuous air space	X	X	X	X	X	X
Air space per floor	X	X	X	X	X	X
Vertical office module wide air spaces	X	X	X	X	X	X
Each office module a separate air space	X	X	X	X	X	X

Double skin façades have several advantages:

- Energy efficient. Many designers share that conviction.
- The outer skin protects the solar shading in between, which favours its service life.
- High inside surface temperatures upgrade comfort. Allows glazing from floor to ceiling.
- Better sound insulation than well glazed single skin façades.

Only a performance evaluation can show if this picture fits with the reality.

8.4.5.2 Building physics: heat, air, moisture

Thermal transmittance

The air space between the two glazed skins functions as a ventilated cavity. In addition, a supply façade allows the enthalpy of the air entering the inside space to intervene in the heat balance at building level, while a curtain flow of inside air extracted along the double skin façade participates in fixing the temperature of the mixed air in the air handling units. Independent of what happens in terms of enthalpy, the dynamic thermal transmittance of a double skin façade follows from:

8.4 Glass façades

$$U_{dyn} = \frac{1}{\dfrac{1}{U_{w,1}} - \dfrac{1}{h_e}} \left[\frac{\theta_i - \dfrac{1}{h}\int_0^h \theta_{c,1}(z)\,dz}{\theta_i - \theta_e} \right] \tag{8.31}$$

with $U_{w,1}$ the thermal transmittance of the inner skin and $\theta_{c,1}$ the surface temperature at the air space side. How the function $\theta_{c,1}(z)$ looks, depends on the complexity of the calculation model used. In their simplest form, the heat balances of both the inner and the outer skin and the air space in between look like:

Inner skin (1)

$$\frac{1}{\dfrac{1}{U_{w,1}} - \dfrac{1}{h_e}}\left(\theta_i - \theta_{c,1}\right) + h_c\left(\theta_c - \theta_{c,1}\right) + h_R\left(\theta_{c,2} - \theta_{c,1}\right) = 0$$

Air space (c)

$$h_c\left(\theta_{c,1} - \theta_c\right) + h_c\left(\theta_{c,2} - \theta_c\right) = c_a\, G_a\, \frac{d\theta_c}{dz} \tag{8.32}$$

Outer skin (2)

$$\frac{1}{\dfrac{1}{U_{w,2}} - \dfrac{1}{h_i}}\left(\theta_e - \theta_{c,2}\right) + h_c\left(\theta_c - \theta_{c,2}\right) + h_R\left(\theta_{c,1} - \theta_{c,2}\right) = 0$$

In the three equations, θ_c is the temperature at the centre plane in the air space. G_a represents the air flow along that space (kg/s). c_a is the specific heat capacity of air (J/(kg·K)), $\theta_{c,2}$ the surface temperature at the outer skin air space side, h_c the convective surface film coefficient and h_R the surface film coefficient for radiation in the air space. The possible boundary conditions are: $z = 0$, $\theta_c = \theta_i$ or $z = 0$, $\theta_c = \theta_e$, depending on whether inside or outside air enters the air space. With forced convection, the air flow G_a is a design variable. With stack induced convection, a flow equation expressing the equilibrium between stack force and hydraulic losses from inlet to outlet, complements equation (8.32).

The balances show the dynamic thermal transmittance of a double skin façade depends on the air flow passing along. That is why the adjective 'dynamic' is used. With outside air, the higher the flow, the more the dynamic value increases compared to the value without flow. Of course, for a supply double skin façade, the air enters the indoor spaces preheated.

The opposite holds for inside air: the higher the flow along, the lower the dynamic thermal transmittance compared to the value without flow (Figure 8.35). For a moderate increase of the dynamic thermal transmittance and hardly any decrease in thermal comfort with outside air passing along, well insulating glass should form the inner skin whereas single glass suffices for the outer skin. The opposite is the case with inside air passing along: single glass for the inner skin and well insulating glass for the outer skin.

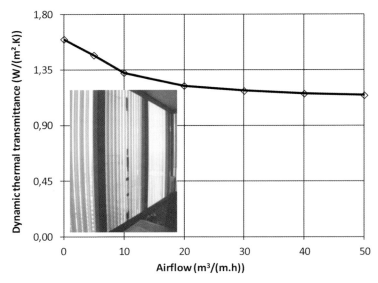

Figure 8.35. Extract double skin façade: outer skin low-e double glass, $U_{gl} = 1.6$ W/(m² · K), 0.154 m wide air space, inner skin single glass, dynamic thermal transmittance as function of extract inside airflow per meter run.

Transient response

As for glass, insolation defines the transient response of a double skin façade. When flowing along the cavity of a sun-radiated façade, the colder outside air heats up. Inside air instead may continue cooling down, although heating prevails during warm days. A transient response may have a decisive impact on thermal comfort!

In outside air washed, in extract and in curtain flow double skin façades, the airflow diminishes some of the gains.

But in general, double skin façades have a solar shading device in the air space. To evaluate shading efficiency, analogous heat balances in terms of the dynamic thermal transmittance intervene, with the shading device as an extra layer and the solar radiation absorbed by each layer as an additional factor. The shading device creates two air spaces that each handle part of the airflow. The solar transmittance finally combines the direct gains by transmission with the indirect gains by long wave radiation and convection at the inside surface of the sun-warmed inner skin. Measured data and calculated results underline solar transmittance of a double skin façade with shading in between passes the value for shading at the outside. Temperatures at the inside surface of the inner skin may also run high as the following case shows.

In a new, 34 storey high-rise office building (Figure 8.36) the envelope consisted of an air curtain type double skin façade, with air flowing bottom-up between the skins. The building was conditioned using chilled ceilings combined with slightly cooled ventilation air. The air entered across ceiling slots, washed the offices, and was extracted at floor level by the double skin façade, where it flowed upwards to return to the central air handling units. That way some 30 m³/(m · h) moved up the façade during working hours. The outside skin consisted of double glass, the inside skin of 6 mm single glass, while the air space in between was 15.4 cm wide and contained a solar shading device made of white lamellas, each turning around a vertical

8.4 Glass façades

Figure 8.36. On the left the office building, on the right a bay of the double skin façade with the solar shading closed.

Table 8.27. Double glazing: properties.

LTA	τ_S	ρ_S	a_S	U
-	-	-	-	W/(m² · K)
0.62	0.38	0.29	0.33	1.6

central axis. For the properties of the double glazing, see Table 8.27. As Figure 8.35 above shows, the dynamic thermal transmittance of the double skin façade reached 1.18 W/(m² · K).

Once the building was in use, employees working in offices facing east, south and west complained about very warm conditions during sunny weather. Apparently, the inside surface of the single glass heated to quite high temperatures, creating a situation where the body gained radiant heat from the façade, while loosing radiant heat to the chilled ceiling.

To diagnose the situation, two bays, one facing south at the 32nd floor and one facing north at the 30st floor, were equipped with thermocouples at the cavity side of the double glass, the inside surface of the single glass and in the middle of the air space outside and inside the solar screen, at 30 cm, 132 cm and 234 cm height (Figure 8.37). Measurements were taken during the summer of 2007, which was neither warm nor very sunny. Meanwhile, a comfort meter registered the situation in an office at the 32nd floor, while a second comfort measurement was taken in a landscape office at the 25th floor with the double skin façades facing east and south.

Figure 8.38 gives the inside surface temperatures of the south looking bay, measured during the warmest week of 2007. Peak values up to 35 °C are logged during working hours for an air temperature in the office nearing 30 °C. Highest temperature noted at the single glass beyond working hours was 47 °C. At this moment, temperatures in the double skin façade looked as shown in Figure 8.39.

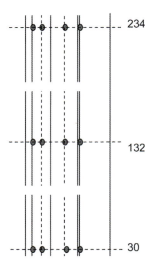

Figure 8.37. Position of the thermocouples in the double skin façade.

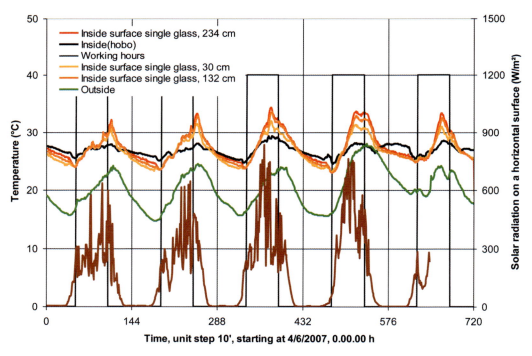

Figure 8.38. Temperatures on the inside surface of the single glass inside skin, air temperature in the office, outside climatic data.

Why such high temperatures? There are three reasons: the outside skin is too transparent for solar heat, its low U-value restraints heat conduction to the outside and the air flow washing the MSF is too limited.

8.4 Glass façades

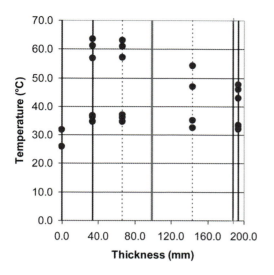

Figure 8.39. Temperatures in the double skin façade. Lowest points are the daily means, highest the peaks.

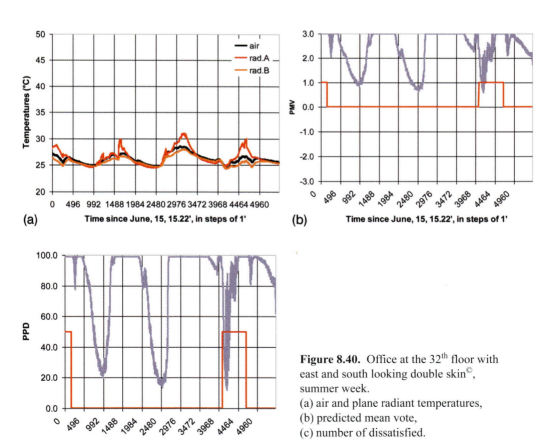

Figure 8.40. Office at the 32th floor with east and south looking double skin$^©$, summer week.
(a) air and plane radiant temperatures,
(b) predicted mean vote,
(c) number of dissatisfied.
Red (0, value) line indicates working hours.

As Figure 8.40 illustrates for the office at the 32th floor, comfort measurements revealed an unacceptable situation. The predicted mean vote for a metabolism of 1.2 Met and summer clothing could go up from cold in the morning (below −1) to 3 in the afternoon, much too warm, ending with 100% dissatisfied. The same was true for the landscape office at the 25th floor. Hence, the complaints of the employees were clearly justified.

Or, double skin façades do not necessarily upgrade thermal comfort. To limit overheating in the case being, the outer skin should consist of double glazing with very low solar transmittance, the solar screen should have a reflective outside surface and a higher return flow should wash the air space. An alternative of course is an inner skin of well insulating glass, a single glass outer skin, a solar screen with reflective outside surface and washing the cavity in and out with outside air.

Moisture tolerance

Few people believe double skin façades cause problems. However, when the inner skin lacks air-tightness, the inside relative humidity is quite high and the outer skin consists of single, sometimes normal double glass, surface condensation against the air space side of the outer skin may occur as the following case shows.

The office building had a ground floor and four upper floors that formed an extension of an existing building. The five floors were arranged around a central atrium. The building had an air curtain type double skin façade at the two corners (Figure 8.41), with the air flowing top down between the skins. The air entered the office spaces through VAV-boxes at floor level, left at ceiling level, was conducted to the top of the active windows, flowed down the air space and returned underneath to the air handling units. The outside pane of the façade consisted of double glass, the inside pane of single glass. The air space in between was 24 cm wide and contained a roller blind as solar shading device. During working hours, the air flow along the cavity varied from 100 to 140 $m^3/(h \cdot m)$. The dynamic thermal transmittance of the double skin façade reached 1.26 $W/(m^2 \cdot K)$.

After the building was occupied, two problems arose: (1) users complained about bad thermal comfort during winter and (2) surface condensation deposited in the façade at the aluminium jambs and double glazing of the outer skin.

The diagnosis was based on comfort measurements and infrared pictures of the façade. Therefore, a comfort meter was installed at an employee's position close to the façade. For a metabolism of 1.2 Met and winter clothing (Clo = 1.2), overall comfort during working hours

Figure 8.41. Building outside view, view from inside on the MSF and schematic section of the MSF.

8.4 Glass façades

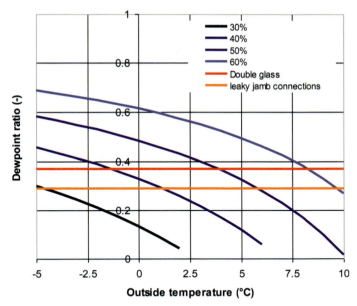

Figure 8.42. Dew point ratio for a given inside relative humidity, compared to the surface temperature ratio at the double glass edge and the aluminium jambs.

oscillated around a PMV of –1.04, a PPD of 28.5%. The draft rate reached 61.2%. A comparison with ASHRAE standard 55-2004 showed these conditions were unacceptable.

The reason was lack of airtightness of the inner skin. In fact, each alternate single glass panel was operable. Near the rebates, air velocities up to 0.5 m/s where measured, which indicated wind-induced leakage of cool air from the double skin into the office.

The outer skin however should not have suffered from surface condensation. In fact, aluminium jambs with thermal break were used having an equivalent thermal transmittance close to the central value of double glass. Of course, the glass edge spacers induced some thermal bridging, resulting in lower surface temperature ratios there than in the central part. A detailed inspection however revealed that the joints between the aluminium rails and jambs in the outer skin were not sealed. This allowed outside air to enter the void aluminium profiles, annihilating the effect of the thermal break over a limited distance. It was also important that the inside relative humidity in winter was kept around 50%, as logging showed

Figure 8.42 depicts the consequences. Each time the outside temperature drops below 6 °C for an inside relative humidity around 50%, surface condensation starts. That outside temperatures below 6 °C are common during the colder months in a moderate climate explains why surface condensation provoked complaints.

8.4.5.3 Building physics: acoustics

At first sight, a comparison with double windows could be made. Operable sashes in the inner skin, however may jeopardise good noise transmission loss, as indoor noise entering the façade gets reflected in the air space between the two skins, spreading out over larger surfaces that way. Avoidance demands breaks in the air space per floor and between neighbour office bays, i.e. a return to a separate air space per office module.

8.4.5.4 Fire safety

Double skin façades complicate fire fighting. A continuous air space allows easy flame spread between floors, while fire fighters have to break two glass surfaces to enter the building. The best solution is to divide the air space into storey-high and office module wide separate volumes.

8.4.5.5 Building level: energy efficiency

A detailed study evaluated: (1) a stack flow, outside air washed double skin façade, called DSF, (2) a forced flow, inside air washed, central air handling unit coupled extract double skin façade, called AFW, (3) a forced flow, outside air washed, central air handling unit coupled supply double skin façade, called SUP for energy efficiency by looking at the net energy demand for heating and cooling in terms of primary energy (chillers consume electricity which has a primary energy multiplier of 2.5 in countries with mainly thermal power plants). None of the three were better than a classic, well insulated curtain wall with high performing outside solar shading. Of the three, AFW performed the best with an increase in net primary energy demand of 4 to 5%. DSF gave an increase of 7 to 9% whereas SUP ended in plus 16 to 18%.

Only a so-called smart double skin façade, which functioned as a SUP during cold weather and switched to AFW when warm, succeeded in limiting the increase to 2.8%. Double skin façades with variable air flow and a software based predictive control may even do better, although due to higher fan consumption the savings in terms of net energy demand may evaporate in terms of end energy use. And, envelope controls still are a delicate, less robust technology.

8.4.6 Photovoltaic façades (PV)

Although costs are dropping steadily, photovoltaic cells remain quite expensive. That is why usage gives the best results in terms of investment costs if the PV replaces an outside finish. For some years, experiments were run using PV as façade and roof cover, turning both surfaces into solar power plants. Proposals such as tiles covered with PV are less appropriate. Proposals such as tiles with adhered PV-cells are less appropriate. The service life of a tile in fact surpasses the service life of PV and few people will exchange the tiles because the PV-cells lose too much efficiency. Important in case of appropriate application is that the cells do not become too hot in sunny weather. If this happens, efficiency drops. PV façades and roofs should therefore apply the ventilated cavity principle, i.e. behind the PV sits an open, permanently outside air washed cavity.

8.5 References and literature

[8.1] Caluwaerts, P., Verougstraete, P. (1979). *k-waarde van vensters*. WTCB-cursussen (in Dutch).

[8.2] WTCB (1980). *Onderzoeksprogramma vernieuwbouw, houten buitenschrijnwerk*. Vorderingsrapport, 26 pp. (in Dutch).

[8.3] DIN 18055 (1981). *Fenster, Fugendurchlässigkeit, Schlagregendichtheit und Mechanische Beanspruchung, Anforderungen und Prüfung* (in German).

[8.4] Werner, H. (1981). *Wärmedurchgang durch Fenster und Wand unter Berücksichtigung der Sonneneinstrahlung*. Haustechnik – Bauphysik – Umwelttechnik, 102. Jahrgang, Heft 3, pp. 121–126 (in German).

8.5 References and literature

[8.5]   Stuvel, P. (1981). *Glaszetters bewijzen zichzelf en producenten bar slechte dienst.* Cobouw Magazine, 7 april, pp. 8–12 (in Dutch).

[8.6]   WTCB (1982). *Rolluiken voor woningen.* Technische voorlichting 143, 52 pp. (in Dutch).

[8.7]   Standaert, P. (1983). *Thermal Evaluation of 4 Types of Windows by the Finite Difference Method.* Programma RD-Energie, DPWB, 12 pp.

[8.8]   Hens, H., Vermeir, G. (1984). *Bouwfysica 4, bouwakoestiek, practische problemen en toepassingen.* ACCO, 193 pp. (in Dutch).

[8.9]   Standaert, P. (1984). *Thermal Evaluation of Window Frames by the Finite Difference Method.* Proceedings of the 'Windows in Building Design and Maintenance', Gotenburg.

[8.10] NBN B62-004 (1987). *Berekening van de warmtedoorgangscoëfficiënt van beglazing* (in Dutch).

[8.11] IEA, Annex 12 (1987). *Thermal transmission through windows.* Final Report, 55 pp.

[8.12] NEN 3661, (1988). *Gevelvullingen, Luchtdoorlatendheid, waterdichtheid, stijfheid en sterkte.* Eisen, NNI (in Dutch).

[8.13] Brandt, J. (1988). *Fensterausbildung.* DBZ 3, pp. 403–412 (in German).

[8.14] Schreiner, H. (1988). *Glasdickendimensionierung.* DBZ 7, pp. 987–989 (in German).

[8.15] ASTM C23 (1989). *Standard test method for measuring the steady state thermal transmittance and conductance of fenestration systems, using hot box methods.*

[8.16] Wilke, G. (1989). *Sanierung alter Holzfenster.* DBZ 7, pp. 913–916 (in German).

[8.17] WTCB (1989). *Glas in daken.* Technische voorlichting 176, 64 pp. (in Dutch).

[8.18] Meert, E. (1989). *Glas in daken, ontwerp en uitvoering.* WTCB-tijdschrift (in Dutch), 2.

[8.19] Janssen, P. (1990). *Structural glazing.* DBZ 7, pp. 1007–1011.

[8.20] Thermische isolatie, enkele practische oplossingen aan de hand van kunststof-isolatiematerialen (1990). BKV, Brussel (in Dutch).

[8.21] Müller, H. (1991). *Fenster und Fassaden.* DBZ 7, pp. 261–264 (in German).

[8.22] Kempenaers, F. (1994). *Thermische optimalisatie van aluminium ramen.* Eindwerk KU Leuven (in Dutch).

[8.23] Wouters, P., Martin, S., Vandaele, L. (1995). *Vensters, bouwfysisch bekeken.* WTCB-tijdschrift, Winter, pp. 10–19 (in Dutch).

[8.24] Salomez, L. (1995). *Waterinfiltraties via het buitenschrijnwerk.* WTCB-tijdschrift, Lente (in Dutch).

[8.25] Mills, E. (1995). *Windows as Luminaires.* IAEEL-Newsletter, 3–4, pp. 11–13.

[8.26] Van Acker, J., Decaesstecker, C. (1996). *Houten buitenschrijnwerk, bescherming en afwerking.* WTCB- tijdschrift, Herfst, pp. 23–31 (in Dutch).

[8.27] TI-KVIV (1996). *Een klare kijk op glas.* Studiedag, 9 October 1996 (in Dutch).

[8.28] Carmody, J., Selkowitz, S., Heschong, L. (1996). *Residential Windows, A Guide to New Technologies and Energy Performance.* W. W. Norton & Company, New York – London.

[8.29] Zhao, Y., Durceija, D., Goss, W. (1997). *Prediction of the multicellular flow regime of natural convection in fenestration glazing cavities.* ASHRAE Transactions, Vol. 103, Part 1.

[8.30] Klems, J., Warner, J. (1997). *Solar Heat Gain Coefficient of Complex Fenestrations with a Venetian Blind for Differing Slat Tilt Angles.* ASHRAE Transactions, Vol. 103, Part 1.

[8.31] Zhao, Y., Curcija, D., Goss, W. (1997). *Prediction of the Multicellular Flow Regime of Natural Convection in Fenestration Glazing Cavities.* ASHRAE Transactions, Vol. 103, Part 1.

[8.32]   Baker, J., Henry, R. (1997). *Determination of Size-Specific U-Factors and Solar Heat Gain Coefficients from Rated Values at Established Sizes-A Simplified Approach*. ASHRAE-Transactions, Vol. 103, Part 1.

[8.33]   Bielefeld, H. (1997). *Doppelfassadentechnik, ein Konstruktionsprinzip zu Energieeinsparung*. DBZ 10, pp. 125–132 (in German).

[8.34]   Haddad, K., Elmahdy, A. (1998). *Comparison of the monthly thermal performance of a conventional window and a supply-air window*. ASHRAE-Transactions, Vol. 104, Part 1.

[8.35]   Ingelaere, B. (1998). *Toepassing van de nieuwe norm EN ISO 717-1:1996, deel 1: akoestische prestaties van glas*. WTCB-tijdschrift, Lente, pp. 24–34 (in Dutch).

[8.36]   Ingelaere, B., Vermeir, G. (1998). *Geluidisolatie van vensters*. WTCB-tijdschrift, Herfst, pp. 26–34 (in Dutch).

[8.37]   Gertis, K. (1999). *Sind neuere Fassadenentwickelungen bauphysikalisch sinnvoll? Teil 2: Glas-Doppelfassaden*. Bauphysik, Vol. 21, pp. 54–66 (in German).

[8.38]   Lyons, P., Arasteh, D., Huizinga, C. (2000). *Window performance for human thermal comfort*. ASHRAE-Transactions, Vol. 105, Part 1.

[8.39]   Baker, P., Saelens, D., Grace, M., Takashi, I. (2000). *Advanced Envelopes*. Final Report IEA, EXCO ECBCS, Annex 32, Uitgeverij ACCO, Leuven, 64 pp. + App.

[8.40]   Saelens, D. (2002). *Energy performance assessment of single storey multiple-skin façades*. Doctoraal Proefschrift KU Leuven.

[8.41]   Saelwns, D., Carmeliet, J., Hens, H. (2003). *Energy performance assessment of multiple skin façades*. International Journal of HVAC&R 9(2), pp. 167–186.

[8.42]   Gustavsen, A., Arasteh, D., Kohler, C., Curcija, D. (2005). *Two-dimensional conduction and CFD simulations of heat transfer in horizontal window frame cavities*. ASHRAE-Transactions, Vol. 111, Part 1.

[8.43]   Anon. (2005). *ASHRAE Handbook of Fundamentals, Chapter 15*. ASHRAE, Atlanta, Georgia, USA.

[8.44]   Jiru, T., Haghighat, F., Perino, M., Zhanhirella, F. (2006). *Zonal modelling of double skin façades*. Proceedings of EPIC 2006 AIVC, Lyon, 20–22 November, pp. 129–134.

[8.45]   Safer, N., Gavan, V., Woloszyn, M., Roux, J. (2006). *Double skin façade with venetian blind: global modelling and assessment of energy performance*. Proceedings of EPIC 2006 AIVC, Lyon, 20–22 November, pp. 229–234.

[8.46]   Blomsterberg, A. (2006). *Best practices guidelines for double skin façades*. Proceedings of EPIC 2006 AIVC, Lyon, 20–22 November, pp. 937–942.

[8.47]   Micono, C., Perino, M., Serra, V., Zanghirella, F., Filippi, M. (2006). *Performance assessment of innovative transparent active envelopes through measurements in test cells*. Research in Building Physics and Building Engineering (Eds.: P. Fazio, H. Ge, J. Rao, G. Desmarais), Taylor & Francis, pp. 293–300.

[8.48]   Lingnell, A., Spetz, J. (2007). *Field Correlation of the Performance of Insulating Glass Units in Buildings–A Twenty-Five Year Study*. Proceedings of Buildings X, Clearwater Beach, December 5–9 (CD-Rom).

[8.49]   Van den Engel, P., Mixoudis, G. (2008). *The development of a climate façade for a hot humid climate*. Proceedings of NSB 2008, Copenhagen, June 16–18, pp. 127–134.

[8.50]   Jensen, R., Kalyanova, O., Heiselberg, P. (2008). *Modelling a naturally ventilated double skin façade with a building thermal simulation model*. Proceedings of NSB 2008, Copenhagen, June 16–18, pp. 143–150.

8.5 References and literature

[8.51] Kalyanova, O., Jensen, R., Heiselberg, P. (2008). *Data set for empirical validation of a double skin façade model*. Proceedings of NSB 2008, Copenhagen, June 16–18, pp. 151–158.

[8.52] Haase, M., Amato, A. (2008). *Controlling ventilated façades*. Proceedings of NSB 2008, Copenhagen, June 16–18, pp. 1159–1166.

[8.53] Anon. (2009). *ASHRAE Handbook of Fundamentals, Chapter 15*. ASHRAE, Atlanta, Georgia, USA.

[8.54] Becker, R. (2009). *Curtain wall leakage-interrelation with construction details*. Energy Efficiency and New Approaches (Eds.: N. T. Bayazit, G. Manioglu, G. K. Oral, Z. Yilmaz), pp. 177–184.

[8.55] Serra, V., Zanghirella, F., Perino, M. (2009). *Experimental energy efficiency assessment of a hybrid ventilated, transparent façade*. Energy Efficiency and New Approaches (Eds.: N. T. Bayazit, G. Manioglu, G. K. Oral, Z. Yilmaz), pp. 247–254.

[8.56] Van Den Bossche, N., Janssens, A., Moens, J. (2009). *Watertightness of window frames: experience of notified bodies*. Energy Efficiency and New Approaches (Eds.: N. T. Bayazit, G. Manioglu, G. K. Oral, Z. Yilmaz), pp. 369–378.

[8.57] Labaki, L., Castro, G., Gutierrez, G., Caram, R. (2009). *Thermal behaviour of window films in glass façades*. Energy Efficiency and New Approaches (Eds.: N. T. Bayazit, G. Manioglu, G. K. Oral, Z. Yilmaz), pp. 379–384.

[8.58] Heim, D., Janicki, M. (2010). *Thermal behaviour and efficiency of ventilated double skin façade in Polish climatic conditions*. Proceedings of CESBP 2010, Cracow, September 13–15, pp. 257–234.

[8.59] Skogstad, H., Sivert Uvsløkk, S. (2010). *Sealing of Window and Door Joints in Timber Frame Buildings and Watertightness*. Proceedings of Buildings XI, Clearwater Beach, December 2–7 (CD-Rom).

[8.60] Teasdale-St-Hilaire, A. (2010). *Innovative Structurally Glazed Curtain Wall at the New Vancouver Convention Centre: Design and Construction Challenges*. Proceedings of Buildings XI, Clearwater Beach, December 2–7 (CD-Rom).

[8.61] Murphy, M., Gustavsen, A., Jelle, B., Haase, M. (2011). *Energy savings potential with electrochromic switchable glazing*. Proceedings of NSB, Tampere, 29 May – 2 June, pp. 1281–1288.

[8.62] Haase, M. (2011). *Heat transfer in ventilated double façades with obstructions, Energy savings potential with electrochromic switchable glazing*. Proceedings of NSB, Tampere, 29 May – 2 June, pp. 1297–1304.

9 Balconies, shafts, chimneys and stairs

9.1 In general

With windows and outer doors mounted, the enclosure is finally rain and windproof. Some specific building parts remain: balconies, chimneys, shafts, and staircases. Balconies vary in area. Shafts come in different forms and dimensions. Chimneys are shafts with a specific function of acting as a smoke flue. Stairs allow, as lifts do, vertical circulation.

The chapter starts with balconies, followed by shafts, chimneys and stairs.

9.2 Balconies

9.2.1 In general

The main function of balconies is to create an 'exterior environment' for people living at the higher floors of medium or high-rise buildings. Besides, they serve as escape route, gangway for maintenance of façade and glazing, etc.

9.2.2 Performance evaluation

The systematic approach, followed for floors, façade systems and roofs, remains the reference although we take more freedom now in what to evaluate. Air tightness, thermal transmittance, transient response and sound insulation do not play a role for balconies. However they may cause annoying thermal bridging. Solving this issue should not jeopardize structural integrity, moisture tolerance and fire safety.

9.2.2.1 Structural integrity

Balconies cantilever. That means their support experiences the highest bending moment with traction at the top and compression at the bottom. If executed in reinforced concrete, the main reinforcement should be located above. This moment of course relieves the floor slabs, although a correct design requires enough counterpoise, which is why it is logical to cast balconies and slabs as a unit. Otherwise, they torque the façade beam.

9.2.2.2 Building physics: heat, air, moisture

The relevant performances are thermal bridging and moisture tolerance.

Thermal bridging

The down side of the structural solution sketched is true thermal bridging. With glass façade and massive outer walls insulated inside, both the linear thermal transmittance (ψ) and the temperature ratio (c_{hi}) fall short. Outside insulation and cavity fill score better because the only disappointment is the linear thermal transmittance (Figure 9.1). Possible upgrades include:

1. Shuttering up the façade beam and the floor with an insulating permanent formwork and mounting some insulation on top of the floor slabs before casting the screed (Figure 9.2).
2. Supporting the balconies at each load bearing partition or façade column by cantilevered beams having an insulating permanent formwork (Figure 9.3).
3. A free standing load bearing structure supporting the balconies (Figure 9.4).
4. Inserting thermal cut strips, whose structure allows withstanding the bending moment between balcony and slab (Figure 9.5).

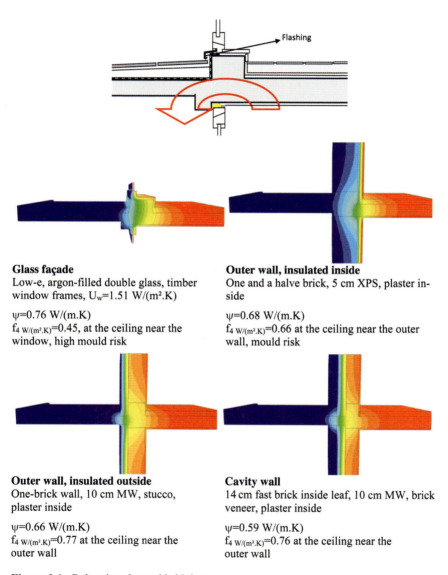

Glass façade
Low-e, argon-filled double glass, timber window frames, U_w=1.51 W/(m².K)

ψ=0.76 W/(m.K)
$f_{4\ W/(m^2.K)}$=0.45, at the ceiling near the window, high mould risk

Outer wall, insulated inside
One and a halve brick, 5 cm XPS, plaster inside

ψ=0.68 W/(m.K)
$f_{4\ W/(m^2.K)}$=0.66 at the ceiling near the outer wall, mould risk

Outer wall, insulated outside
One-brick wall, 10 cm MW, stucco, plaster inside

ψ=0.66 W/(m.K)
$f_{4\ W/(m^2.K)}$=0.77 at the ceiling near the outer wall

Cavity wall
14 cm fast brick inside leaf, 10 cm MW, brick veneer, plaster inside

ψ=0.59 W/(m.K)
$f_{4\ W/(m^2.K)}$=0.76 at the ceiling near the outer wall

Figure 9.1. Balconies, thermal bridging.

9.2 Balconies

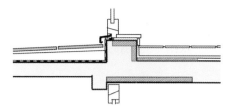

Figure 9.2. Balcony, insulating permanent formwork, insulation on top of the floor slab.

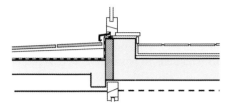

Figure 9.3. Balcony supported by cantilevered beams.

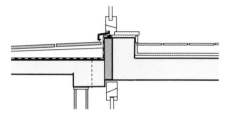

Figure 9.4. Balcony, separate load bearing Structure.

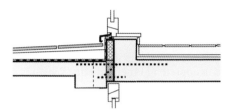

Figure 9.5. Balcony, using thermal cut elements.

Upgrade (1) is not only inefficient, the solution also increases the thermal load of façade beams and floor slabs. Upgrade (2) allows thermal cuts between balcony and floor coplanar with the façade insulation. They of course should stay rainproof and interstitial condensation free, while the choice presumes a careful check on strength and stiffness of the supporting beams. Upgrade (3) excels as again a correct thermal cut between balcony and floor is possible. Of course, rain leakage and harmful interstitial condensation must be excluded and the supporting structure has to stay stable against horizontal load. The lack of co-planarity with the façade insulation of the thermal break could be a drawback. Upgrade (4) only functions when the thermal cut includes the necessary bending and shear force reinforcement. The bars of course degrade the thermal performance somewhat. To avoid corrosion, rain proofing and a control on interstitial condensation are necessary. This is why, depending on the microclimate expected, galvanized steel (inland climates), normal stainless steel (salt intrusion possible) or special stainless steel bars (coastal regions) must be used.

Moisture tolerance

A problem with balconies is rainwater seepage. When finished with tiles, rain penetration along the joints saturates the sand or screed underlay, resulting in tile heaving and rot of membranes with felt or jute insert. Without membrane, long lasting seepage ends in stalactites near shrinkage cracks in the balcony slab, stalagmites on the floor below and bar corrosion, which in turn spalls concrete coverage and balcony edges.

9.2.2.3 Durability

Balconies experience large temperature differences between top and underside with linked bending facilitating crack formation in the concrete slab. Therefore, insulating the top surface before waterproofing makes sense, although thermal load on screed and finish then increases. One solution is to finish balconies with loose laid tiles on stilts.

9.2.2.4 Fire safety

With balconies the 1 meter path length rule between the windows at successive floors is easily met. In fact, a 0.5 meter wide balcony suffices. In hospitals, continuous balconies wide enough for sickbeds to pass serve as evacuation paths.

9.2.2.5 User's safety

The user's safety prevails. Each balcony must have handrails. Their height must exceed the centre of gravity of a standing adult, which means more than 1.2 m. The design should also prevent children from climbing over and wriggling between the posts.

9.2.3 Design and execution

The short analysis shows that when designing and constructing balconies, four details must be considered: (1) thermal cut, (2) water proofing and drainage, (3) finish, (4) handrails.

9.2.3.1 Thermal cuts

The factory made thermal cuts advanced as upgrade (4) combine preformed high density EPS strips with galvanized or stainless steel tension reinforcement, compression bars and 45° bent shear bars, see Figure 9.6. Sometimes contractors construct such thermal cuts using normal steel bars. Corrosion risk makes this an unacceptable practice.

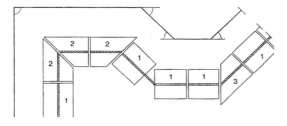 

Figure 9.6. Preformed thermal cut blocks.

9.2.3.2 Water proofing and drainage

On top of the concrete, a correctly built cantilevered balcony gets a polymer bitumen membrane that is pulled up at the perimeter some 15 cm above the finish where it gets a protective flashing (see Figure 9.1). 3 to 4 cm thick, punch resisting XPS boards with loose-laid tiles on stilts as finish come on top of the waterproofing. Rainwater discharges along the edges of small balconies. Larger balconies direct the rain at membrane level towards downspouts. Due to the limited height of cantilevered slabs, traps are located beneath the fall pipes.

If balconies cover inside spaces, execution resembles low-sloped roof construction with loose-laid tiles on stilts above the membrane. Both the membrane and the insulation must of course withstand the pressure the stilts exert without punching or excessive deformation. Stresses in the insulation may not pass 1/3 of its compressive strength.

9.2.3.3 Floor finish

As mentioned, preference goes to loose-laid stony or timber tiled floors on stilts. If that is not possible, a drainage layer above the membrane is provided, which discharges into the downspouts, and the balcony is finished with a tiled floor divided in up to 10 m² large bays with resiliently sealed expansion joints in between and along the façade.

9.2.3.4 Hand rails

Handrails should withstand and transmit a horizontal load of 1000 N/m, the working point 1.2 m above the floor finish, to the balcony slab. Massive handrails consist of masonry or reinforced concrete, both with discharge openings at drainage plane level. Masonry failing to transmit the 1000 N/m needs reinforcement. Lightweight handrails have vertical posts coupled by horizontal bars. The posts are bending proof fixed at the front or the underside of the balcony slabs. How to finish is a matter of choice (Figure 9.7).

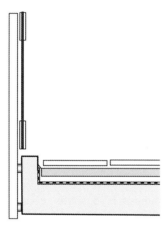

Figure 9.7. Balcony, finished lightweight handrail.

9.3 Shafts

9.3.1 In general

Shafts exist in any form and dimension. Lifts need purpose-designed shafts, in former times closed, now often glazed. Buildings with extended networks of pipes and ducts need vertical shafts passing from floor-to-floor. Their location and section needs thorough consideration, surely when flexibility is a prime requirement. Per floor, horizontal distribution of pipes and ducts above hung ceilings or below raised floors starts at these shafts. This requires a floor-to-floor height of 3.6 m or more.

9.3.2 Performance evaluation

Structural integrity, air-tightness, sound insulation, and fire safety are important.

9.3.2.1 Structural integrity

Lift- and duct shafts help ensure structural response against horizontal loads, which is why in mid- and high-rise buildings they are assembled in and form part of the central core.

9.3.2.2 Building physics

Air tightness

Shafts couple all floors. If air leaky, thermal stack will develop along the building height, in winter under-pressurizing the lower and over-pressurizing the higher floors. The consequences are outside air infiltration at the lower and inside air exfiltration at the higher floors, worse sound insulation between the inside and outside at the lower floors, higher mould and surface condensation risk at the higher floors, too much ventilation by infiltration at the lower floors, bad ventilation at the higher floors, smells from below invading the higher floors, and annoying whistling noises when lift doors close. Once closed, lift and duct shafts should therefore be as airtight as possible. This requirement gains in urgency with building height.

Acoustics

Sound transmission loss by shafts must equal the values advanced for partitions and floors separating the spaces the shafts couple: typically $R_w(C) > 52$ dB. Closing all floor passages in pipe and duct shafts with cast concrete or sprayed foam may ensure this. When using concrete, pipes and ducts must be wrapped with a mineral wool first to allow movement once the passage is filled. Walls and doors of lift shafts must have the right sound transmission loss, a possibility, which may replace a floor passage closure in duct shafts. If serving as partition between zones, each enclosing shaft wall should get half the sound transmission loss required.

9.3.2.3 Fire safety

Malfunctioning in case of fire directly relates to a lack of air tightness. If so, shafts serve as oxygen supply and paths for fire spread. Avoidance demands excellent air-tightness plus walls and passage closures with sufficient fire resistance (60 of 90′).

9.3.3 Design and execution

Reinforced concrete and masonry lift shafts have walls thick enough to ensure the required sound insulation. An alternative are shafts consisting of a steel skeleton filled with safety glass that run along the façades or in glass covered patios. Each stop has a door bay with double set of airtight closing sliding doors, wide enough for people in wheelchairs to pass. Shafts have to be wide or deep enough to house the lift car and the rails with counterweight. The bottom space contains shock absorbers and sometimes the electric motor and lift mechanism. Lift rooms at the top only house motor and mechanism. The pneumatic lifting jack is located under hydraulic lifts. In case of fire alarm, only fire fighters can use the lifts.

Also larger pipe and duct shafts are constructed in reinforced concrete or brick, with walls that ensure air tightness, sound insulation, and fire safety. Small shafts instead have a board encasement. Filling the floor passages with concrete or sprayed foam should guarantee air-tightness, sound insulation, and fire safety in such cases.

9.4 Chimneys

9.4.1 In general

The performances expected from chimneys differ largely from other building parts. That is why we look to design considerations and execution related information only.

9.4.2 Design considerations

The chimney section determines smoke flow rate. Additional parameters are smoke temperature, the height from connection to exhaust, and chimney course. Flow rate and smoke temperature depend on stove or boiler type and capacity. Height and course are design data. The driving forces are thermal stack, which temperature and chimney height define, and, if present, fan power. Take temperature first. The heat balance per infinitesimal small chimney height dz equals (Figure 9.8):

$$d\Phi = U_{chimney} \left(\theta_s - \theta_o \right) dz \tag{9.1}$$

with θ_o temperature around, θ_s smoke temperature and $U_{chimney}$ thermal transmittance of the chimney per meter run (W/(m · K)), the result of convection between smoke and the chimney's inside surface, conduction across the chimney's side walls and long wave radiation plus convection between the chimney's outside surface and the environment. The better a chimney is insulated, the lower heat loss dΦ. Determining specific thermal transmittance demands a two dimensional calculation or is done experimentally. Heat loss to the environment now lowers smoke enthalpy:

$$dH_s = c_s \, G_s \, d\theta_s \tag{9.2}$$

where G_s is smoke flow rate in kg/s and c_s smoke specific heat capacity, hardly different from air. Heat loss equal to enthalpy loss yields:

$$U_{chimney} \left(\theta_s - \theta_o \right) = c_s \, G_s \, \frac{d\theta_s}{dz} \tag{9.3}$$

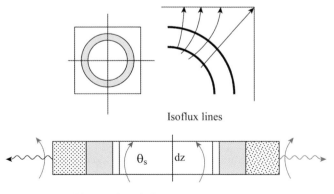

Figure 9.8. Chimney: heat balance.

with as solution:

$$\theta_s = \theta_o + (\theta_{s,o} - \theta_o) \exp\left(\frac{-U_{chimney}\, z}{c_s\, G_s}\right).$$

$\theta_{s,o}$ being smoke temperature at the chimney entrance. The ratio $c_s\, G_s / U_{chimney}$ is called the chimney constant $C'_{chimney}$. The larger this constant is, the warmer the smoke and the stronger thermal stack. Its value increases with smoke flow rate and better chimney insulation.
Thermal stack then becomes:

$$\Delta p_T = 3640 \int_0^{H_{stack}} \left[\frac{1}{T_s(z)} - \frac{1}{T_0(z)}\right] dz$$

with H_{stack} the height between fire seat in the boiler and the chimney's exhaust.
The smoke velocity (v_s) follows from the equilibrium between thermal stack and pressure loss along the path 'from the outside across the air inlet grid to the boiler room, through the boiler room to the boiler and from within the boiler along the chimney to the outside':

$$\Delta p_T = f\, \frac{L_{stack}}{d_{H,stack}}\, \frac{\rho_s\, v_s^2}{2} + \sum\left(\xi_j\, \frac{\rho_s\, v_s^2}{2}\right) + Z_b + Z_g \qquad (9.4)$$

L_{stack} in this equation represents the unwind length and $d_{H,stack}$ the hydraulic diameter of the chimney, $\Sigma(\xi_j)$ the sum of all local resistances between boiler connection and chimney exhaust, Z_b the smoke related hydraulic resistance of the boiler and Z_g the hydraulic resistance of the ventilation grid in the boiler room. Underpressure in the boiler is subtracted from stack while overpressure or boiler fan pressure adds. (9.4) can also be written as:

$$A_{stack} = \left(f\, \frac{L_{stack}}{2\, \rho_{rg}\, d_{H,stack}} + \frac{\Sigma \xi_j}{2\, \rho_s} + \frac{Z_b + Z_g}{\rho_s\, v_s^2}\right)^{0.5} \frac{G_s}{\sqrt{\Delta p_T}} \qquad (9.5)$$

with Δp_T the stack corrected for under-, over- or boiler fan pressure. The given equations allow optimizing chimneys in terms of section, unwind length, height, and thermal insulation. Too few stack or a too small section in fact diminishes transportable smoke flow, which in turn lowers air supply to the stove or boiler, a main cause of incomplete combustion, CO production and less available capacity.
Of course, the equations are not simple. They not only demand iteration but hydraulic friction and local resistances are only approximately known. This hinders an exact determination of chimney temperature, which is why one rewrites Equation (9.4) for smaller capacities as:

$$A_{stack} = \frac{C_1\, G_s}{\sqrt{\Delta p_T}} \qquad (9.6)$$

with C_1 a coefficient, which depends on the hydraulic resistances, the unwind chimney length, the chimney section, etc. Thermal stack equation in turn is simplified to:

$$\Delta p_T = C_2\, H_{stack} \qquad (9.7)$$

9.4 Chimneys

wherein C_2 depends on the temperatures in and around the chimney. The following relation couples flow rate G_s in kg/h to stove or boiler capacity Φ_b in kW:

$$G_s = 2.6\, \Phi_b \tag{9.8}$$

Implementing (9.7) and (9.8) in (9.6) and writing $C_1/\sqrt{C_2}$ as $1/n$, gives:

$$A_{stack} = \frac{2.6\, \Phi_b}{n\sqrt{H_{stack}}} \tag{9.9}$$

an equation known under as 'Redtenbacher's equation', which links capacity to chimney height and section. The coefficient n differs between fuels and chimney types. Its value increases with better chimney insulation and higher smoke inlet temperature. It drops with increasing unwind length, more curves, a smaller section, the use of chimney pipes with higher roughness, etc. With stove or boiler capacity in kW and chimney section in m², n moves between 900 and 1800.

9.4.3 Design and execution

9.4.3.1 Stoves

Wood, coal, and oil stoves demand chimneys with a height of at least 4 m till smoke exhaust. Table 9.1 gives the sections. For prefabricated chimneys, the values reduce by 25%.
Stoves burning natural gas actually get concentric air intake/smoke outlet to the outside.

Table 9.1. Chimneys for stoves: section.

Coal and oil stoves		Section	
Capacity kW	Connections allowed	Brick laid	Composed if circular prefabricated elements
≤ 18	2 at the most	13.5 × 13.5 cm	13.5 cm
12 to 30	2 to 4	13.5 × 20 cm	16.5 cm
24 to 48	4 to 8	20 × 20 cm	20 cm

9.4.3.2 Boilers

For the section, we refer to Table 9.2, which is based on Redtenbacher's equation. In former times chimneys were brick laid. Today, they are made up of an inner and outer pipe with a temperature proof thermal insulation in between. When in concrete, the prefabricated elements are mounted on each other. Steel pipes instead form one prefabricated piece. Chimneys for large capacity boilers demand a design based on the full method. In such cases inner and outer pipe are typically made of stainless steel or reinforced concrete.

Quite new are chimneys that act as heat exchanger between air supply and smoke exhaust. The elements consist of a smoke pipe and a concentric air supply pipe whose outer surface is insulated and encapsulated. The sucked outside air reaches the boiler in counter flow with the smoke, while being heated to the smoke inlet temperature. This increases the efficiency of boilers, while the boiler's air demand decouples from the indoor environment, a fact excluding CO-emission indoors. Of course, both the smoke and air pipe must be gas tight.

Table 9.2. Boiler chimneys, section.

Section (A_{stack})		Height (H_{stack}) in m				
Rectangular cm × cm	Circular ϕ in cm	≤ 10	10–15	15–20	20–25	25–30
		Boiler capacity (Φ_b) in kW				
10/10	10	≤ 10				
13.5/13.5	13.5	10–18				
13.5/20	16.5	12–30				
20/20	20	24–48				
	23	58	≤ 64			
20/22		48–71	94	113		
	26	81	93	105	110	
20/25		97	126	152	176	197
	30	128	145	163	187	209
25/30		164	213	255	293	328
	37	192	221	244	279	291
30/35		251	324	387	445	498
	45	291	349	372	419	442
40/40		425	546	652	748	837
	52		488	547	582	640
40/50			719	858	985	1102
	60		698	768	837	896
50/60			1185	1415	1624	1817
60/60			1484	1772	2034	2277
65/65				2160	2480	2777

9.5 Stairs

9.5.1 In general

As lifts, stairs allow vertical circulation. In case of fire, they also figure as escape routes. This introduces requirements in terms of width. Concerning form and layout, a distinction exists between (Figure 9.9):

1. Straight-run stairs
2. Straight-run stairs, divided in two parts by an intermediate landing
3. Angular stairs in two parts with intermediate landing
4. Opposite running stairs in two parts with intermediate landing
5. Twofold angular stairs in three parts with two intermediate landings
6. Opposite running stairs in three parts with two intermediate landings
7. Single flight stairs with winding steps in the upper part. Figure 9.10 shows how to wind

9.5 Stairs

8. Single flight stairs with winding steps in the lower part (there are also single flight stairs with winding steps in the lower and upper part)
9. Opposite running stairs with wheeled steps
10. Spiral stairs
11. Solid newel stairs

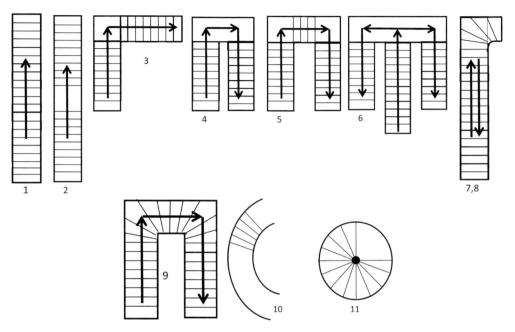

Figure 9.9. Different stair forms. The numbers relate to text above.

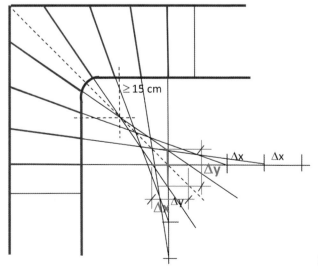

Figure 9.10. Winding the stair steps.

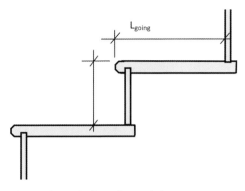

Figure 9.11. Stair: going and rise.

Steps have a going and rise (Figure 9.11). Following relation should be respected between going depth and rise height:

$$L_{going} + 2\, H_{rise} = 64 \quad [cm] \tag{9.10}$$

The length in stepping direction of an intermediate landing should obey:

$$L_{landing} = 64 + L_{going} \quad [cm] \tag{9.11}$$

A going much longer than the rise height gives an easy stair. The opposite produces a steep stair. Easy stairs have a heading function in the circulation patterns as is the case at building entrances or for helical staircases at ground floor level. Emergency stairs are typically steep.

Stairs are made of reinforced concrete, timber, or steel. They cantilever from a load bearing wall or span freely between floors or intermediate landings. Opposite running stairs in two parts sometimes have floating intermediate steps.

9.5.2 Performance evaluation

Again, we dispense with the systematic discussion. Stairs have to be structurally safe. Insulation against contact noise, fire resistance and user's safety are also important.

9.5.2.1 Structural integrity

In case of cantilevered stairs, the stair slab is calculated as being restrained in the load-bearing wall. If a stair spans freely between floors, it acts as a beam under axial load and bending (Figure 9.12). Opposite running stairs in two parts with a floating intermediate step experience a more complex set of forces. On the one hand each part bends, on the other hand the whole assembly experiences tension, compression, and torsion.

9.5 Stairs

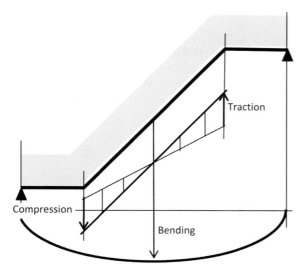

Figure 9.12. Free spanning stair, axial load and bending.

9.5.2.2 Building physics: acoustics

Insulation against airborne noise

Not the stair but the stair hall is the noisy space with high reverberation time, which is why for example all partition walls between hall and apartments should show high noise transmission losses ($R_w(C) > 55$ dB).

Isolation against contact noise

The stair itself is the problem now. A solution is resilient supports (Figure 9.13).

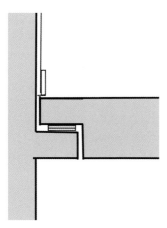

Figure 9.13. Stair, resilient support.

9.5.2.3 Fire safety

As already stated, stairs act as escape routes. The hall they are in therefore demands treatment as a fire compartment with walls and floors ensuring a fire resistance beyond 90′ and a smoke exhaust at roof level. The stair itself is built of a non-burnable material guaranteeing a fire resistance beyond > 30′. Moreover, the expected number of people to evacuate defines step width. For hospitals, the requirements for example are:

Number of people n	Step width in m Coming down	Number of people n	Step width in m Going up
$0 \leq n \leq 96$	1.2	$0 \leq n \leq 60$	1.2
> 96	$1.25\, n/100$	> 60	$2\, n/100$

9.5.2.4 User's safety

Stairs must have handrails. Their height must equal 80 cm, while the design should be such that children can neither climb over nor wriggle between the posts. Steps should also be rough to avoid slipping.

9.5.3 Design and execution

We refer to the literature. Stair construction may take many directions. The design in fact belongs to global architecture. Reinforced concrete stairs are poured on site or prefabricated. For single flight stairs, this is done per part, for spiral and solid newel stairs per step. Steel stairs are manufactured and mounted on site. The same holds for timber stairs.

9.6 References and literature

[9.1] Koninklijk besluit tot vaststelling van de normen inzake beveiliging tegen brand en paniek waaraan ziekenhuizen moeten voldoen (1980). Belgisch staatsblad, pp. 458–503 (in Dutch).

[9.2] Frick, Knöll, Neumann, Weinbrenner (1988). *Baukonstruktionslehre, Teil 2*. B. G. Teubner Stuttgart (in German).

[9.3] Cziesielski, E. (1990). *Lehrbuch der Hochbaukonstruktionen*. B. G. Teubner Stuttgart (in German).

[9.4] WTCB (1995). *Balkons*. Technische Voorlichting 196, 96 pp. (in Dutch).

[9.5] WTCB (1995). *Houten trappen*. Technische Voorlichting 198, 107 pp. (in Dutch).

[9.6] Brandschutz in öffentlichen und privatwirtschaftlichen Gebäuden (1996). Bertelsmann Fachzeitschriften-Supplement, 100 pp. (in German).

[9.7] Anon. (1996). *Good practice guide 174, Minimising thermal bridging in new dwellings*. Department of the Environment, U. K.

[9.8] Anon. (1996). *Good practice guide 183, Minimising thermal bridging when upgrading existing dwellings*. Department of the Environment, U. K.

[9.9] Neufert (2009). *Bauentwurfslehre*. 39th ed., Vieweg + Teubner Verlag, 550 pp. (in German).

10 Partitions; wall, floor and ceiling finishes; inside carpentry

10.1 Overview

Once the building fabric is ready and the enclosure or part of it is wind and rainproof, completion starts with casting top floors where possible, fixing non-bearing lightweight partitions and installing all building services, included heating, cooling, ventilation, air conditioning, domestic cold and hot water, waste water discharge, distribution of gasses and liquids, electricity, data infrastructure, lifts, moving stairways. Next, come the remaining top floors, followed by the finishing of hung ceilings and surfaces. Painting and decorating complete the building.

How the inside finishes look is largely the architect's responsibility. We limit the discussion to partitions walls, top floors, hung ceilings, and doors. Finishes are hardly treated.

Completion work includes its own problems. Fitting ducts still requires a lot of chopping and breaking, generating a lot of debris, while plastering and casting top floors produce significant quantities of building moisture.

10.2 Partition walls

10.2.1 In general

Partition walls are classified as light- and heavyweight. Brick laid or cast concrete ones are part of the carcass work, even when non-bearing (Figure 10.1). If load-bearing, their thickness equals or exceeds 14 cm. If non-bearing 9 cm suffices. Although in timber framed construction load-bearing partitions are lightweight, in brick and concrete buildings lightweight means non-bearing, either removable or non-removable.

Dismountable and removable partitions consist of separate modules with everything that goes with it. Adjusting screws allow clamping them between screed and ceiling, whereupon electricity wiring is drawn across purpose-provided hollow sections for connection to built-in plug sockets and switches. Fixing non-removable partitions starts with mounting a metallic stud and transverse skeleton, after which all bays are filled with mineral wool and both sides

Figure 10.1. Heavy-weight partition wall, masonry.

Performance Based Building Design 2. From Timber-framed Construction to Partition Walls.
First Edition. Hugo Hens.
© 2013 Ernst & Sohn GmbH & Co. KG. Published 2013 by Ernst & Sohn GmbH & Co. KG.

are covered with gypsum or any other screwed board material. Finishing lightweight partitions includes mounting doorposts and plinths, hanging the doors, painting, and fixing plug socket and switch cappings.

10.2.2 Performance evaluation

10.2.2.1 Structural integrity

Load bearing partitions belong to the building's structural system. As such, they carry a large part of the useful and dead floor loads, part of the slab weights and their own weight down to the foundations. They also help resist all horizontal loads. Dimensioning considers axial load and bending, even when vertical compression alone seems to intervene. Unavoidable eccentricities and potential buckling are the reasons why (Figure 10.2). Due to uncertainty, load reduction factors or the consideration of allowable stresses are preferred design aids.

Non-bearing partitions must carry their own weight per storey and have adequate resistance against horizontal impacts. They must also allow hanging up some heavy objects and have surfaces with correct hardness once finished.

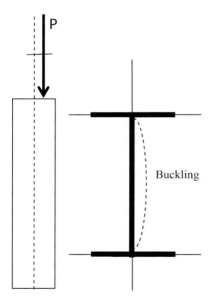

Figure 10.2. Load bearing partitions: load eccentricity and buckling.

10.2.2.2 Building physics: heat, air, moisture

Air tightness

Air tightness does not seem a critical performance, except for party walls. Acceptable sound insulation in fact demands air tightness. When the connection with the outer wall cavity stays unclosed, outside air may wash cavity party walls, causing unexpected heat losses and degrading the building's transmission transfer coefficient. Also timber framed walls can suffer from outside air washing, which is why the connection with the outer walls demands correct tightening, see Chapter 1.

10.2 Partition walls

Thermal transmittance

In some countries, legal requirements exist for party walls between homes and apartments, in Belgium:

	U-value $\leq$ W/(m² · K)
Party wall between homes or apartments	1.0

A cavity party wall without ties with a 2 cm thick mineral wool or glass fibre cavity fill meets this requirement. Of course, principals can impose project specific requirements. With low energy buildings for example, the current practice requires a whole wall thermal transmittance below 0.5 W/(m² · K) for party walls.

Transient response

As has been mentioned, six parameters define the transient of a space, zone or room: (1) glazing (type, surface, orientation, slope), (2) presence or absence of solar shading (how, where), (3) ventilation schedule, (4) surface and thermal storage capacity of the partitions and floors, (5) surface and thermal inertia of the opaque façade assemblies, (6) internal gains. Thermal storage capacity of a partition wall depends on weight and finish. Heavyweight partitions help in damping transient response, on condition however they are not wrapped with insulation. Of course, thermal storage capacity is not the sole argument in choosing between light and heavyweight. Others are flexibility, weight restrictions, costs, etc.

Moisture tolerance

Heavyweight partitions in new constructions contain building moisture. Drying must proceed within a reasonable period without damage. A specific problem with partitions walls is sucking floor cleaning water. To avoid that, masonry walls get a waterproof layer inserted some 15 cm above floor slab level. Gypsum block walls in turn are wrapped at the bottom with a waterproof foil, whereas lightweight walls mounted on the top floor have bitumen-felt strips underneath (Figure 10.3).

Figure 10.3. Partitions walls, combating cleaning water suction: on the left for masonry walls, in the middle for gypsum block walls, on the right for lightweight walls.

Thermal bridging

Heavy-weight partitions cause thermal bridging when located on floors on grade or on floors, which separate indoors from outdoors, when both have the insulation located between slab and screed (Figure 10.4).

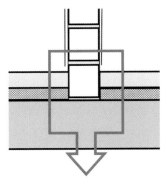

Figure 10.4. Heavyweight partitions acting as thermal bridges.

This is a reason why preferring insulation underneath the slab, while only the load bearing partition walls act as thermal bridge at ground floor level then. For their part, envelopes insulated at the inside change all junctions with heavyweight partitions into thermal bridges.

10.2.2.3 Building physics: acoustics

Necessary sound insulation depends on the partition wall's function. Party walls between homes or apartments require high acoustical quality, with an $R_{w,500}$-value preferentially above 60 dB. Cavity walls without ties, 14 cm thick leaves, uncoupled floors and the acoustical cut passing across the foundation comply (Figure 10.5). For partition walls in the same home or apartment, the requirements are less strict. In office buildings, the partitions enclosing meeting rooms and management offices demand sufficient sound transmission loss.

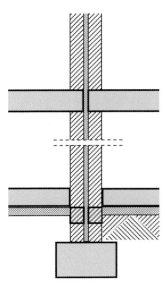

Figure 10.5. Cavity party wall without ties.

10.2 Partition walls

10.2.2.4 Durability

Drying of building moisture in newly constructed heavyweight partitions may induce shrinkage cracks. If that happens after rendering, the cracks will stand out in the plaster and finish.

10.2.2.5 Fire safety

Again, function establishes the requirements. Partitions separating neighbour fire compartments need a fire resistance beyond 90' in terms of structural safety, smoke tightness and fire spread by radiation.

10.2.3 Design and execution

We limit the discussion to the lightweight partitions. In most cases, the dimensions are based on a multiple of a module M, M being for example 10 cm. Elements with length 12M and height 24M are often used. Door elements have a 9 M wide door bay with an adjacent 3M long adaptation panel. For sickrooms, door bays take the whole 12M. When opting for movable partitions, free divisibility of the floor area and the use of a modular floor raster are of great importance. The possibilities are a 3M × 3M raster with the partitions crosswise on the raster lines, or a more complex (3M + 1M) × (3M + 1M) strip raster with the partitions filling the 1M strips. The raster must repeat itself in the hung ceilings.

Acceptable airborne sound insulation demands lightweight partitions constructed as two-leaf systems, airtight, both leafs flexible and of different thickness, the distance in between as large as possible and the cavity filled with mineral wool. The connections between leaves and metal studs must have limited stiffness, while the contact between metal top plate and ceiling and metal bottom plate and top floor should be resilient and airtight. All this also contributes to fire resistance. Additionally, both leaves must consist of non-combustible material of class A. Sprayed gypsum and gypsum board meet this requirement (Figure 10.6).

Figure 10.6. Example of an acoustically well performing light-weight partition wall.
The skeleton consists of open web metal studs.
Sprayed gypsum on rib mesh forms the leaves, while the cavity in-between is filled with glass fibre.

10.3 Building services

We refer to the literature treating the subject. To be noted: the consequences in terms of space needed, acoustical requirements, and accessibility for control and maintenance of all components must be fully considered and accounted for at the design stage.

10.4 Wall finishes

10.4.1 In general

Most wall and ceiling finishes are done with wet rendering or dry surfacing using gypsum board. Spray gypsum plaster, ready mixed lime plaster, putty plaster and decorative plaster are the wet renders most used. The first two apply in thicknesses between a few millimetres and two centimetres (Figure 10.7). The thickness of putty or decorative plaster is limited to a few millimetres. After spraying or spreading, the plaster has to harden for a while, after which the stucco worker equalizes the surface with a rule.

In case of a dry finish, the stucco worker bonds the gypsum boards to the masonry using adhesive gypsum. As stated above, finishing timber and steel framed lightweight partitions is done by screwing gypsum boards against the studs, or, for outer walls, against the battens forming the service cavity. Once mounted, all sunken joints between boards are reinforced with glass fibre fabric, after which the stucco worker equalizes the whole with putty plaster. For wet plaster as well as for gypsum board, painting or wallpapering completes the finish.

Figure 10.7. Partition wall, wet finish with gypsum plaster.

10.4.2 Performance evaluation

Only the most important requirements are briefly discussed

10.4.2.1 Structural integrity

Aside from sufficient resistance against bumping and punching and a correct hardness, at first glance there are no structural requirements for inside finish. At first glance. The finish gives the wall the necessary air tightness. Bonding strength at outer walls therefore must be high enough to transmit most of the wind pressure structurally active substrate:

10.4 Wall finishes

Bonding strength	N/mm²
Minimum value	0.1
Wished value	0.2

10.4.2.2 Building physics: heat, air, moisture

Air tightness

As has been stated, it is the inside finish that makes masonry outer walls acceptably airtight. Wet finishes perform better from that point of view than a dry one. One reason is that dry finishes are always left with a narrow air layer between board and wall, which functions as a route for outside air to penetrate the voids in the connected lightweight partition walls.

Looking to timber framed construction, both a non-penetrated gypsum board finish and the air and vapour retarding foil at the inside of the insulation guarantee air tightness. However, also the connections between outer walls and inside partitions must be constructed to exclude air penetration. Leave a service cavity on the inside of the foil and once all wiring and ducting are in place fill it with dense mineral wool before fixing the gypsum boards.

We discussed the positive consequences of good air tightness in previous chapters. It is important to know that each perforation of an airtight finish or foil creates air leakage. If for example timber framed outer walls lack an air and vapour retarding foil with service cavity, all plug sockets and switches will act as air leaks.

Moisture tolerance

At first glance, moisture tolerance of inside plasters is hardly a concern. With gypsum plaster, two facts nevertheless demands consideration: sucking floor cleaning water and short-circuiting the waterproof layer down all masonry walls. Due to the capillarity of gypsum plaster it is best to stop at this waterproof layer and apply a water repellent cement plaster underneath.

10.4.2.3 Building physics: acoustics

If an inside plaster hardly adds weight, its air tightness allows the mass law to play its full role once masonry walls are rendered. For timber-framed construction, the acoustical merits of gypsum board are even more pronounced. Together with the insulation and the outside OSB sheathing, it creates a composite wall, which takes maximal advantage of the difference in weight and stiffness between both leaves and the sound absorption by the mineral wool.

10.4.2.4 Fire safety

Gypsum plaster and gypsum board are non-combustible materials. They also retard temperature increase during fire for a while. In fact, bounded water in the gypsum evaporates at 100 °C, transforming sensible into latent heat, which tempers temperature.

10.4.3 Design and execution

10.4.3.1 Wet plasters

When plastering, sharp edges have to be reinforced using galvanized steel rim profiles. Substrates that hardly allow bonding are first covered with a strengthening fabric. Gypsum causes

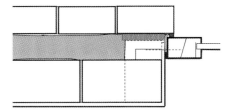

Figure 10.8. Masonry cavity wall: a putty joint guarantees continuity and air tightness between plaster and window (the blue line visualizes the airtight plane).

steel to corrode, which is why embedded metal parts require protection. Connecting joints between plaster, windows, and outer doorposts require air tightness (Figure 10.8). Painting is done once the plaster is dry. Quality control consists in measuring levelness, orthogonality and squareness and evaluating hardness. Requirements are:

What?	Requirement?
Levelness	Deviations less than 4 mm/m
Orthogonality	Deviation less than $\dfrac{\sqrt[3]{h_{\text{wall}}}}{8}$ with h_{wall} height of the wall
Squareness	Less than 5 mm for an area with length up to 2 m
	Less than 3 mm for an area with length above 2 m

10.4.3.2 Gypsum board

In cases where gypsum board acts as air retarding layer, all connecting joints with windows, doorposts, ceilings, and floors require airtight sealing.

10.5 Floor finishes

10.5.1 In general

Finishing floors may include a levelling out layer, one or more separation layers, a top flooring or screed and the floor cover. For the last, tiles, parquet, PUR resin, and flexible coverings such as carpets, linoleum, vinyl and others are used. In case a raised floor applies, the propping system stands directly on a floor deck levelling out layer, which is needed only when the finished floor includes separation layers or is raised. The term 'separation layer' encompasses the thermal insulation, the resilient layer in case of a floating floor and/or the waterproof membranes needed. Thickness of the levelling out layer depends on the ducting package laying on the slab (Figure 10.9) as it must embed all of them to offer a flat surface ready for further finishing. Raised floor systems demand an especially high level of flatness.

The top flooring or screed brings the floor package at the desired level for final finish, while offering the flatness and acoustical performances required. We distinguish bonded screeds from non-bonded screeds, floating floors and screeds for floor heating. The first lie directly on the slab or blind floor, the second have a separation layer in between, the third upgrade contact noise reduction and the fourth cover or contain the floor heating pipes (Figure 10.10).

10.5 Floor finishes

Figure 10.9. Floor slab with plastic draw-in pipes on it.

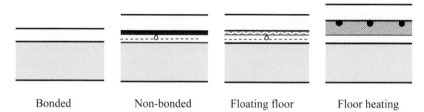

Bonded   Non-bonded   Floating floor   Floor heating

Figure 10.10. The different screeds.

Screeds consist of a mixture of sand and cement or anhydrite. The last is used in self-levelling screeds. Sometimes lightweight aggregates such as cork granules, EPS-pearls, perlite granules, vermiculite granules are added. There are also dry screeds. They consist of purpose-adapted gypsum board sheets glued together.

The way conduits and draw-in pipes could be built-in inspired developments, among others of prefabricated screed elements, which have cut-outs in two orthogonal directions at their underside.

Raised floors take a separate position. They include a propping system spread on a modular, quadratic raster adjusted to the right height. Then come load bearing floor tiles in high-density particleboard, which are finished with carpet or any other flexible floor cover. In the space between levelling layer and tiles come all cable bundles for electricity and data transmission. They connect to building blocks in the tiles, containing the necessary plug sockets. The main advantage of raised floors is that they allow adapting and changing wiring for electricity and data transmission at any moment, which is a necessity now in office buildings.

10.5.2 Performance evaluation

Only the most important requirements are briefly discussed

10.5.2.1 Structural integrity

Aside from sufficient hardness, punching and bonding strength, there are no other requirements for bonded screeds. With non-bonded screeds, floating floors and screeds for floor heating, due to the screed functioning as a slab on a resilient substrate, useful load, dead load of the floor cover and own weight causes bending moments and shear forces. Bending gives a peak moment around each local load and dying, alternating positive and negative moments further away from the load. To withstand the changing moment patterns, the screed is reinforced with a welded 38 × 38 × 1 mm to 150 × 150 × 4 mm steel mesh, mounted single or double. Single means the mesh comes mid-plane. Double indicates one mesh on top and one underneath.

With raised floors, the particleboard tiles must transmit the useful load and the own weight to the propping system without too much bending and vibrating.

10.5.2.2 Building physics: heat, air, moisture

Air tightness

Air tightness must be carefully considered when finishing floors above crawlspaces. Air ingress from a wet crawlspace across leaks left around pipes in the finished floor may add quite some water vapour to the inside air and could be the cause of mould problems at thermal bridges in the enclosure in cold and moderate climates, see Figure 10.11.

For upper floors, air tightness is normally not an issue. Experience shows raised floors deserve special attention. When for cost reasons the outer walls are not plastered between the levelled slab and raised floor, the enclosure may show manifest air leakage. The under floor then serves as a plenum for infiltrating outside air, which enters the offices via the plug sockets in the raised floor. The result is persisting draft complaints by the employees.

Figure 10.11. Mould on the lintel (on the left) caused by the ventilation pipe connecting the living room to the crawlspace (the pipe is where the lamp sits, see picture on the right).

Thermal transmittance

Insulation and screed guarantee the legally imposed thermal transmittances, for example:

Part	U-value $\leq$ W/(m$^2 \cdot$ K)	R-value surface to surface m$^2 \cdot$ K/W
Floors above grade	0.3 (ISO EN 13370)	1.75
Floors above basements or crawlspaces	0.3 (ISO EN 13370)	1.75
Floors separating indoors from outdoors	0.3	
Floors between apartments	1.0	

10.5 Floor finishes

For a thermal transmittance of 1 W/(m² · K), a lightweight screed or an onsite sprayed cellular concrete screed suffices. Of course, between apartments, floating floors are preferred. A surface to surface thermal resistance of 1.75 m² · K/W requires an insulation layer.

Transient response

A bonded, heavyweight screed with non-insulating floor cover, such as dark coloured tiles, gives optimal thermal storage capacity.

Moisture tolerance

One problem with newly cast screeds is drying time. Figure 10.12 expresses how long it takes, even when a windproof new construction is heated, when air dryers are installed, and fans facilitate surface evaporation. With parquet or any vapour retarding flexible floor cover, execution must wait until the mean moisture ratio in a sand/cement screed drops below 2.5 % kg/kg. For anhydrite, this value is 0.6 % kg/kg. With stony covers, execution can start at an average moisture ratio below 5 % kg/kg for sand/cement, and 1% kg/kg for anhydrite screeds.

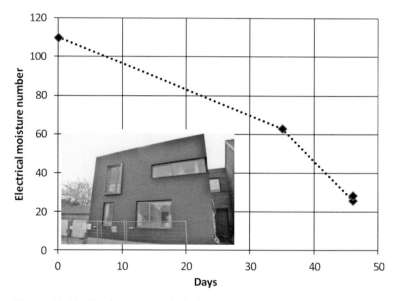

Figure 10.12. Newly cast screed, drying.

10.5.2.3 Building physics: acoustics

Screeds help in upgrading the acoustical performance. In general, they add extra mass. If correctly executed, floating floors provide a significant increase in contact noise attenuation (Figure 10.13). For more details, see the chapter on floors in Performance Based Building Design 1.

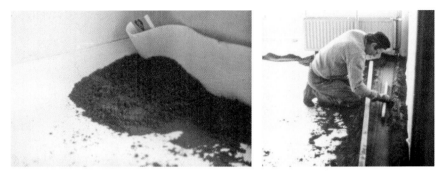

Figure 10.13. Floating floor.

10.5.2.4 Durability

Screeds shrink. To avoid capricious cracking, joints are needed all along the screed's perimeter, while complex floor surfaces require prolongation of some of these joints to form rectangular floor fields. By doing so, one should avoid floor fields that exceed 40 m² and have lengths and widths above 8 m.

10.5.3 Design and execution

Anhydrite screeds lack moisture tolerance. Complete drying and no humidification afterwards are therefore very important. Pipes passing finished floors must be directed through feed-through fittings with the space between fitting and pipe filled with mineral wool and closed up and down with resilient putty. Also shrinkage joints in the screed are filled with soft material and finished with resilient putty. Joints crossing the floor surface should have edges, reinforced with metal strips at finish level, see Figure 10.14.

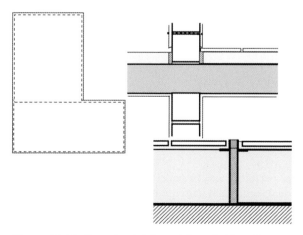

Figure 10.14. Screeds: shrinkage joints, finish at floor cover level.

Screeds are controlled for surface cohesion and flatness with the following requirements:

Flatness	Requirement?	
	1 m lath	2 m lath
Floor class 3	Deviation below 5 mm/m	Deviation below 6 mm/(2 m)
Floor class 2	Deviation below 3 mm/m	Deviation below 4 mm/(2 m)
Floor class 1	Deviation below 2 mm/m	Deviation below 2 mm/(2 m)

Class 1 applies for stony floors, class 2 for all other floor covers, and class 3 refers to the area close to outer walls and partitions. If required, witness screed tiles are cast first and tested for compression strength and resistance against punching. When necessary also thermal and hygric properties are measured.

10.6 Ceiling finishes

10.6.1 In general

Ceiling finishes include wet plastering and the dry alternative of hung ceilings, which includes board as well as lamella solutions (Figure 10.15). Both are mounted closed or open underneath or hung at the floor slab. Closed means one disconnects the ceiling plenum from the room below. Open hung ceilings instead figure as sound absorbing surfaces. The section across a storey with raised floor and hung ceiling is shown in Figure 10.16.

Today, lamella hung ceilings perform additional functions. Chilled ceilings for example have gained in popularity. In such case the hollow lamellas contain chilled water pipes that connect at the edges to a supply and a return.

Figure 10.15. Closed gypsum board hung ceiling.

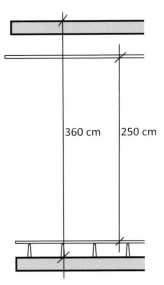

Figure 10.16. Section of a storey with raised floor and hung ceiling.

10.6.2 Performance evaluation

10.6.2.1 Structural integrity

The ceiling structure must be strong enough to bear its own weight and the weight of other components, such as light fittings, chilled beams and air inlets.

10.6.2.2 Building physics: heat, air, moisture

Air tightness

Air tightness is critical only when the airborne and contact noise insulation requires a closed hung ceiling. Then, air leaks must be avoided.

Thermal transmittance

Normally no one expects the ceiling finish to aid in the delivery of the required thermal transmittance. However, in order to improve sound absorption in the plenum, mineral wool bats are placed on top of airtight hung ceilings. These also act as thermal insulation. If in such cases the ceiling complements a low-sloped roof, the bats form an inside insulation layer with related consequences such as worse transient response, larger temperature variations in the load bearing roof slab, high interstitial condensation risk at the underside of this slab, unexpected thermal bridging at the roof edges, etc. Especially for indoor climate class 4 and 5 buildings, these consequences are annoying. With indoor swimming pools, some designers saw this combination of absorption and thermal insulation as a cheap alternative. The consequences were disastrous.

Let it be noted that top to bottom an indoor swimming pool roof should look as follows:

- A compact roof assembly with vapour barrier of class E3/E4 between insulation, which sits directly under the membrane, and screed (or deck if no screed is needed)

10.6 Ceiling finishes

- An intensely inside air washed plenum between deck and hung ceiling
- The open ceiling acoustically absorbing with 'open' taken literally. The ceiling must be designed in a way to activate convection of inside air into the plenum

Transient response

A closed hung ceiling disconnects the thermal storage capacity of any heavyweight deck from inside. If one wishes to activate this capacity, the only way is to use strongly fractionated solutions, which only act as a sound absorber. In the last ten years thermally activated concrete slabs emerged as new trend in exploiting floor capacity. Again, the combination with sound absorption demands strongly fractionated hung ceilings.

10.6.2.3 Building physics: acoustics

As already mentioned, hung ceilings have several functions acoustically. When closed, they upgrade airborne sound transmission loss and contact noise attenuation. In fact, together with the finished floor the hung ceiling forms a composite assembly that functions as a double leaf wall. In auditoriums and concert halls, closed hung ceilings figure as sound reflectors. If open, their main function is increasing sound absorption, lowering reverberation time and heightening speech intelligibility this way.

A problem arises when sound absorbing hung ceilings are combined with movable lightweight partitions. From a mobility point of view, preference goes to a continuous hung ceiling in contact with the movable partitions. However this results in huge sound leaks. A solution consists of covering the hung ceiling above the gridlines, where partitions could be located, with airtight flexible boards that have a sound absorbing layer on top. Mounting fixed vertical cross pieces between a hung ceiling and floor at these gridlines figures as an alternative.

10.6.2.4 Fire safety

Hung ceilings may not facilitate flame spread. They should also moderate heating of the floor slab above, increasing fire resistance this way. In order to function accordingly, the hung ceilings must be airtight and non- or or hardly combustible (class A or B of the EN-standard).

10.6.3 Design and execution

Hung ceilings are a modular concept. Application demands designs based on a line or strip raster. The basic module of any hung ceiling is 3M × 3M or 4M × 4M, with fitting pieces at the outer and partition walls.

Open hung ceilings form easy to mount manufactured systems, with adjustable hangers and horizontal main and secondary rails. The main rails click directly at the hangers, the secondary rails at the main ones. Underneath come modular panels or lamellas, which are clicked at the secondary rails. The systems include all provisions to include light fittings, sprinklers, supply grids, chilled beams and others. Air ducts, VAV boxes and pipes sit in the plenum between hung ceiling and deck. Closed hung ceilings are constructed on site. They mostly consist of gypsum board sheets, screwed on a raster of timber battens or galvanized steel sections. A correct levelling of both types is done using lasers.

10.7 Inside carpentry

10.7.1 In general

The term 'inside carpentry' encompasses inside doors as well as built-in cupboards, shafts, and indoor windows. While the last three are mainly made to order, inside doors evolved as an industrial product, which obeys well defined performance requirements, included dimensional coordination and allowed fittings in length, width, height, squareness and curvature. A door construction consists of the doorpost and the door leaf. The doorpost provides the connection with the partition wall and forms the back fillet. Its construction should allow fittings. A revolving door hangs from hinges on the doorpost. The leaf consists of a stiffening edge frame and a core of cardboard honeycomb, particleboard, or thermal insulation, finished at both sides with a sheet of veneer. A door handle, which is connected to a lock case, allows opening and closing. When locking, the nipples pull into the reinforced hook-in places of the doorpost (Figure 10.17). Apart from revolving doors, there are sliding doors, folding doors, etc.

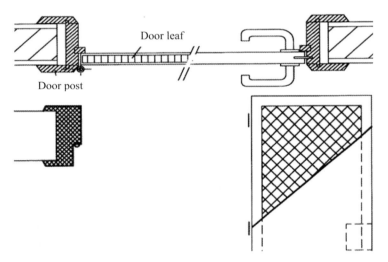

Figure 10.17. Inside door, timber and metal doorpost, door leaf.

10.7.2 Performance evaluation

10.7.2.1 Structural integrity

Structural requirements concern the door leaf, the door fittings, and the doorpost. The last must be constructed and fixed so that repeated door manipulation does not result in premature failure. The door leaf in turn has to show enough resistance against impacts, while the hinges may not come off under the tension and compression forces. The door handle should not become loose with repeated use. Testing consists of loading an open door at its tail end with a vertical force of 1000 N, measuring the deformations, and comparing them with the requirements. Moreover, the door is opened and closed n times, with n defined by the standard (for example 10^6). Afterwards, one measures the deformation and damage and compares with the values allowed.

10.7.2.2 Building physics: heat, air, moisture

With inside doors, air tightness prevails as it defines sound insulation and fire resistance. For doors separating carports from inhabitable spaces, air tightness also avoids polluted air from being sucked into these spaces. The rebate between door leaf and doorpost, included the joint between leaf and sill, and possible joints between doorpost and partition wall form the main leaks. Although the length of the rebate is set equal to the perimeter of the leaf, the sill mostly lacks a rebate. Upgrading air-tightness therefore demands a rebate at that location and rebate strips all along the perimeter. For outer doors, a two-step rebate, together with filling the joints between doorpost and outer walls with PUR-foam or mineral fibre and closing them at both sides with resilient putty, is a good choice (Figure 10.18).

Of course, when the law requires a purpose designed ventilation system with flow through openings in the inside doors, upgrading the rebates makes no sense.

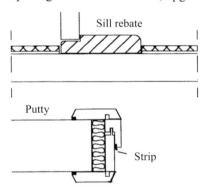

Figure 10.18. Inside doors, caring for air-tightness.

10.7.2.3 Building physics: acoustics

A normal inside door weighing some 10 kg/m² has a sound transmission loss not exceeding 15 dB. Requirements advanced by some standards are:

Application		R_{500} (dB)	
	Minimum	Optimal	
Apartments	Entrance door	27	37
Hotel rooms	Room door	32	37
Sickrooms	Room door	27	37
Class rooms	Door between class and corridor	27	–

If better performances are required, first the ensemble of doorpost and closed door should be as airtight as possible. Besides, a more massive door leaf is needed, for example one where a particle board core is located. In fact noise transmission loss of door leafs obeys the mass law:

$$R_m = 19.8 \log_{10} m - 0.25 \qquad (10.1)$$

Doubling the weight adds some 6 dB to the loss. Constructing the door leaf as a composite assembly 'sheet of veneer/mineral wool/sheet of veneer' is another possibility. But if for any reason doors require a noise transmission loss beyond 30 dB, the only choice left is a two-door solution, though even then the whole must be airtight and both door leaves heavy enough.

Figure 10.19. Sound absorbing flow through vent.

In case the law requires a purpose designed ventilation system with flow through openings in the doors, reconciliation with acceptable noise transmission loss presumes sound absorbing flow through vents (Figure 10.19).

10.7.2.4 Fire safety

Between fire compartments in a building, one needs doors with a 30 or 60′ fire resistance. This presumes doorposts in massive timber or steel, the joints between post and partition walls filled with mineral wool, and massive door leafs. In case the leaf needs transparency, fire-resisting glass is used. The rebates get a strip, which foams when burning (Figure 10.20).

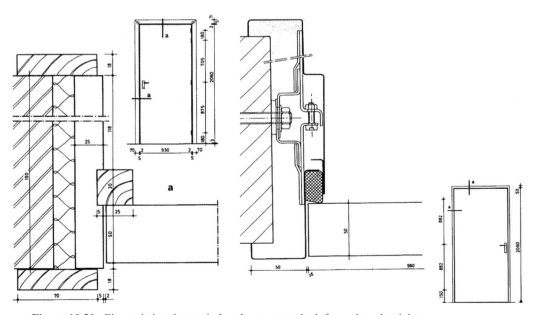

Figure 10.20. Fire resisting doors, timber doorpost on the left, steel on the right.

10.7.3 Design and execution

As stated, doors are industrial products. Their assembly is clarified above. Doorposts are steel or timber based. Positioning the steel ones is part of the rough work. If load bearing, they replace the lintels. Hanging the door leaves demands the post should stay vertically and square and be well fastened into the partition walls. For example one can fix timber posts with special door screws that fit in wall dowels drilled at the height of each hinge.

For more details, reference is made to the literature.

10.8 References and literature

[10.1] Brandbeveiliging/akoestiek en de stukadoor (1977). *Mebest.* 11^e jaargang, No. 5 (in Dutch).

[10.2] Frick, Knöll, Neumann, Weinbrenner (1988). *Baukonstruktionslehre, Teil 2.* B. G. Teubner Stuttgart (in German).

[10.3] WTCB (1992). *Leidraad voor de goede uitvoering van brandwerend binnenschrijnwerk.* Technische Voorlichting 185, 32 pp. (in Dutch).

[10.4] WTCB (1993). *Dekvloeren.* Technische Voorlichting 189, 84 pp. (in Dutch).

[10.5] WTCB (1994). *Dekvloeren, deel 2, uitvoering.* Technische Voorlichting 193, 60 pp. (in Dutch).

[10.6] WTCB (1996). *Binnenbepleisteringen, deel 1.* Technische Voorlichting 199, 52 pp. (in Dutch).

[10.7] WTCB (1996). *Binnenbepleisteringen, deel 2.* uitvoering, Technische Voorlichting 201, 48 pp. (in Dutch).

[10.8] WTCB (1999). *Binnenvloeren van natuursteen.* Technische Voorlichting 213, 52 pp. (in Dutch).

[10.9] WTCB (2000). *Houten vloerbedekkingen: plankenvloeren, parketten en houtfineervloeren.* Technische Voorlichting 218, 154 pp. (in Dutch).

[10.10] Laboratorium Bouwfysica, several reports about damage cases (in Dutch).

[10.11] Manufacturers documentation.

11 Risk analysis

11.1 In general

In both volumes on performance based building design we applied rigorous performance metrics to the design and construction of buildings and building assemblies. In each chapter the link between performances, design and execution was shown. However, designers and builders cannot go very far with performances alone. At the end of the day, they have to come up with a correct design, while realization requires a precise description of how to build. An interesting question to pose is if all solutions that meet the metrics are equivalent in practice. The answer is 'it depends'. Solution x in fact can be more suitable than solution y, if only because of the likelihood of problems occurring afterwards is lower. Risk analysis helps to choose.

11.2 Risk definition

Risk is defined as:

$$\text{Risk} = p \cdot \text{Size} \tag{11.1}$$

with p the probability deficiencies will emerge and size severity of the consequences. A high risk may apply as well to a less likely deficiency with severe consequences, as to a very likely deficiency with less severe consequences. Whether a consequence is classed as 'severe' or not, depends on the reaction of individuals or the society. Defects which cause death are considered very severe, whereas a deficiency that only impairs durability or functioning is less severe. Whereas failing structural integrity might end in collapse, building physics related shortcomings mainly have consequences which degrade usability, comfort, service life, and sustainability. The risk they end in collapse and death toll is rather low, though not impossible as the situation Figure 11.1 shows.

Figure 11.1. In-door swimming pool, low-sloped roof with timber deck, one of the main beams collapsed due to wood rot at the support.

Individuals often perceive severity differently from society. They worry about draft, rain penetration, moisture, and mould. On the other hand, high energy consumption or more CO_2 emissions do not keep them awake, except when prices peak. Of course, there are exceptions. Societies meanwhile, represented by governments, are more concerned about primary energy consumed and related global warming emissions.

11.3 Performing a risk analysis

A risk analysis combines three steps: (1) identification and probability of deficiencies (2) severity of related negative consequences and (3) proposals to limit risk.

11.3.1 Identification and probability of deficiencies

The first step starts with listing all possible deficiencies, followed by a definition of their probability (p). Sometimes probability is determined on the basis of experience. Other sources are damage statistics. Often, deficiencies, like workmanship inaccuracies are given the same probability. Probability may also be derived from measurement experience, as is the case with air tightness. Solutions that should give a guarantee prove to be air leaky once built. Realizing an air permeance coefficient below 10^{-5} m³/(m²·s·Pa) for example appears very demanding, built values easily exceeding 10^{-4} m³/(m²·s·Pa). The reasons are perforated air barriers, overlaps between barrier foils not taped, leaky connections between air barrier and windows, etc.

11.3.2 Severity of the consequences

Next, we identify the severity of each consequence. During design, only simulation can produce the consequences each deficiency generates. These are compared with permissibility criteria and classified according to severity in an operation transposing severity into numbers, for example on a 0 to 100 scale with 0 for 'no problem' and 100 for 'severest'.

11.3.3 Proposals to limit risks

How to select the safest upgrade again requires simulation. A given design may hide n potential deficiencies with probability p_i. Each can produce one or more negative consequences with severity of $Size_{ij}$. Total risk of any upgrade then equals:

$$\text{Risk}_T = \sum_{i=1}^{n} \left(p_i \cdot \sum_{j=1}^{m} \text{Size}_{ij} \right) \tag{11.2}$$

The objective now is to upgrade the design to diminish total risk, the best solution being the proposal producing minimal risk. This is done by eliminating those deficiencies that are most probable and produce the worst consequences, i.e. for which the term between brackets in Equation (11.2) is largest.

11.4 Example of risk analysis: cavity walls

11.4.1 Generalities

In the countries on the North Sea, cavity walls are a common enclosure assembly. Until 1973, they lacked insulation. After 1973, partially or fully insulated cavities became normal practice. Full fills had to take over the cavity function, i.e. acting as capillary break, allowing the veneer's backside to act as drainage plane and eliminating air pressure difference across the veneer.

Before filling started, cavity walls were praised for their moisture tolerance, even when not built perfectly. Thermal transmittance – 1.3 to 1.5 W/(m^2·K) – however was surely too high after the second energy crisis of 1979. Also inside surface temperature in winter dropped excessively, with fast soiling of the inside finish as a consequence. However, with filling, complaints about rain penetration and mould increased. Full fills got most of the blame.

11.4.2 Deficiencies encountered

We differentiate between design errors and workmanship inaccuracies. The brick veneer is assumed carefully laid, and fraud, for instance by not filling the cavity, is excluded. Experiences at building sites in the early 1980s showed this could be too optimistic. Be that as it may, with both assumptions in mind, the following six design errors and nine workmanship inaccuracies were very common in the early 1990s (Figure 11.2):

	Error
D1	Cavity around window and door bays closed, lintels insulated at the inside
D2	Concrete floor slabs contacting the veneer wall
D3	No inside plaster, concrete block inside leaf
D4	Cavity tray missing on the drawings, in the best case mentioned in the specifications
D5	Cavity wall two or more floors high, no tray at each floor
D6	Cavity wall on top not finished with a verge trim
	Inaccuracy
W1	Cavity tray lacking or wrongly mounted (sloping to the inside, hardly any flashing height at the inside leaf)
W2	Cavity tray short-circuited by mortar droppings
W3	Partial fill not lining up with the inside leaf, starts above the tray, neither touching the verge trim nor the roof insulation
W4	Full fill, cavity ties sloping to the inside leaf
W5	Full fill, mortar droppings in the joints between the boards
W6	Full fill, yawning joints between the boards, lower board touching the veneer wall
W7	Full fill, tray above window bays draining sideway on the fill
W8	Partial and full fill, lintels lacking insulation below the tray
W9	Head joints in the inside leaf hardly mortared

Figure 11.2. Design errors and workmanship inaccuracies: some examples.

11.4.3 Probabilities

The literature contains hardly any numbers on the likelihood of the errors and inaccuracies mentioned above. Therefore we had to bring our own building site experience to bear, see Table 11.1. In fairness we must say that thanks to the many in-service trainings for architects, contractors, and site controllers, the situation has been upgraded since the 1990s.

Table 11.1. Probability of the design errors and workmanship inaccuracies.

n^r	Experience
D1	Rather exceptional, ±1 out of 10.
D2	More exceptional, ±1 out of 20.
D3	Really exceptional, ±1 out of 50.
D4	Normal practice. Usually, the contractor corrects the error and puts up the cavity trays. So: 0 on 10.
D5	Normal practice with two-storey buildings (±50% of the dwellings), so, 5 out of 10.
D6	Typical for today's styling, but also traditional styles may suffer from that error (5/10?).
W1	±1 out of 4.
W2	The exception, ±1 out of 10.
W3	Widespread as a problem in Belgium, 95 out of 100. Inside scaffolding is still too common.
W4	Surely half of the ties are mounted that way, each full fill struggles with it (100 out of 100).
W5	Widely spread as a problem in Belgium. ±100 out of 100 full fills suffer from it.
W6	See W5.
W7	Widely spread problem in Belgium. 9 out of 10 full fills show this shortcoming.
W8	Experience showed 5 out of 10 new construction fail at this point.
W9	Widely spread problem in Belgium. Bad filling was enhanced since 14 cm high fast bricks are used for the inside leaf. So, 100 out of 100.

11.4.4 Severity of the consequences

Possible consequences are: (1) worse thermal comfort and draught complaints in winter, (2) insufficient thermal performance and higher heating bills, (3) rain penetration, (4) soiling and more likely mould growth at the inside surface, (5) unexpected high interstitial condensation deposit at the backside of the veneer with salt efflorescence, algae growth and frost damage as a result.

Rain penetration and mould growth cause more stress for owners and inhabitants than high heating bills. The reasons are straightforward: rain seepage and mould are visible. The tangible evidence that 'something is wrong' and the psychological pressure this generates remains as long as the problem stays unsolved. Mould is also perceived as sign of an unhealthy inside environment, even though this is exaggerated according to the scientific facts and figures. Higher heating costs instead are neither visible nor physically perceptible. And inhabitants lack a reference. They do not know what the heating costs would be if the cavity walls were correctly built and filled. In the 1990s, heating costs were also too low compared to the overall living costs to draw attention. That has changed a little today because heating costs have increased in the meantime.

Comfort and draught problems are somewhere in the middle. When they occur they cause concern, but the days that the wind blows so hard and the temperature is so low that draught problems arise are relatively few in moderate climates. Extensive interstitial condensation deposit, due to inside air outflow, humidifying the veneer, is also an in-between case. A non-professional cannot differentiate it from rain absorption. Both wet the veneer. As long as it does not cause damage, a wet surface looks normal.

This allows the compilation of a 'table of discontent' 11.2.

Table 11.2. Weighting the negative consequences.

Order	Negative consequence	Discontentment	Score
1.	Mould	Very high	95
2.	Rain penetration	Very high	95
3.	Interstitial condensation	Low	10
4.	Comfort and draught	Very low	5
5.	Higher energy consumption	Extremely low	0 to 1

The table translates uneasiness in a score, suggesting how many individuals out of one hundred will feel stress. A value of 95 means almost everyone has difficulties with rain penetration and mould. A value of 10 indicates a minority only will complain about salt efflorescence and algae growth due to abundant interstitial condensation. 5 indicates a few will be dissatisfied by the rare moments of draught and discomfort experienced at home, whereas 1 indicates hardly anyone will complain about heating bill increase the deficiencies cause.

The reader of course could question why the number of dissatisfied and not the repair costs figure as a weighting factor. The reason is that all errors and inaccuracies require close to the same costs, which diminishes their usefulness as a weighting factor. The scores reflect the number of court cases. In the 1990s no inhabitant or house owner ever initiated a law suit because of the perception that energy consumption was too high. Instead, mould and rain ended in multiple court cases with sometimes excessive compensations demanded for loss of wellbeing.

Table 11.3 links the negative consequences to the design errors and workmanship inaccuracies. Whereas the six errors affect them all, workmanship accuracies only influence three of them. Two are ranked highest in terms of discomfort. This underlines once again the importance of correct workmanship, though it is up to the designer to control the buildability of her or his proposals.

Table 11.3. Link between errors, inaccuracies, and the negative consequences.

Negative consequence		Error or inaccuracy
D1	Mould	D1, D2, W3, W8, W9
D1	Rain penetration	D3, D4, D5, D6, W1, W2, W4, W5, W6, W7
D2	Interstitial condensation	D3
W3	Comfort and draught	D3
W8	Higher energy consumption	D1, D2, W3, W5, W6, W8, W9

11.4.5 Risk?

At first glance, equation (11.1) looks straightforward enough. Once the probability per error and inaccuracy is known and the consequences scaled – here the number of dissatisfied –, the risk follows by multiplying the two. In reality, things are more complicated.

11.4.5.1 Mould

Whether mould turns into a problem, not only depends on the errors and inaccuracies listed but also on stochastic variables such as the weather, ventilation rate, inside temperatures and vapour release indoors. Where this leads is best judged by looking at measured inside climate conditions. Based on hundreds of weekly means, the 5 and 25% percentiles for temperature and vapour pressure indoors have been determined and related temperature ratios for the coldest month calculated:

$$f_{hi} = \frac{\theta_{si,min} - \theta_e}{\theta_i - \theta_e} \tag{11.3}$$

with $\theta_{si,min}$ the lowest monthly mean temperature at the inside surface of a cavity wall, θ_i the monthly mean reference temperature inside and θ_e the monthly mean reference temperature outdoors. See Table 11.4 for the moderate climate of north-western Europe.

Table 11.4. Inside climate, measured data.

Coldest month at Uccle		Inside				
Temp. °C	Vapour pressure Pa	Zone	Percentile	Temp. °C	Vapour pressure Pa	Temp. factor >
1.7	587 (85% RH)	Day	25	20	1169	0.60
			2.5		1411	0.75
		Night	25	13.5	885	0.58
			2.5		1065	0.81

If we want to limit mould risk to one of four dwellings, lowest temperature ratio may not drop below 0.58. A value above 0.81 limits the risk to one out of forty. Table 11.5 gives the lowest temperature ratio expected looking to the errors D1 and D2 and the inaccuracies W3 and W8. None of the results drops below 0.58. Although the two errors and two inaccuracies are significant, only one in five dwellings will see mould growth developing.

One should exercise caution when using these figures. An infrared picture of a dwelling with partially filled cavity walls suffering from inaccuracy W9 revealed a temperature factor in the corners between two outer walls under 0.2, creating mould risk in seven out of ten homes! Apparently, calculating the thermal bridge without considering the likely airflow patterns produces too high temperature ratios.

Table 11.5. Lowest temperature ratio, looking to the errors D1 and D2, and, the inaccuracies W3 and W9.

Error or inaccuracy		Lowest temperature factor
D1	Cavity closed around window bays	0.67
D1	Lintels insulated inside	0.62
D2	Concrete floor decks touching the veneer	0.63
W3	Partial fill not lining up with the inside leaf	0.63
W8	Lintel not insulated below the cavity tray	0.75

11.4.5.2 Rain penetration

Rain penetration troubles inhabitants. The opinion that complete filling was and is the reason cast the technique in such a negative light that application dropped to one out of twenty homes. However, even with a badly mounted full fill, the probability is not one. To begin with, the wall must face northwest over south to southeast (in Northwest Europe, 95% of the driving rain comes from that direction) and should be exposed. Both conditions are exceptionally fulfilled with terraced dwellings. Not so for detached houses, which in some countries count for 25% of the annual new construction. There we may accept that all have a façade with such or a slightly different orientation. That way, rain penetration becomes a possible unwanted consequence in ±1 dwelling out of 80. However the probability that we will see inaccuracies W4, W5 and W6 (full fill, cavity ties sloping to the inside leaf, full fill, mortar droppings in the joints between the boards, full fill, joints between the boards yawning, lowest board each time touching the veneer wall) in one is 100%. Even then, rain penetration is not certain. Only when the cavity wall suffers additionally from errors D5 and D6 (cavity wall two or more floors high without tray per floor, cavity wall lacking a verge trim on top), does rain penetration risk with wet spots on the inside plaster peak. If insulation thickness does not exceed 6 cm – the case with 80% of the full fills in the 1990s – risk further increases. When the contractor does not correct error D4 (cavity tray missing on the drawings) or the inaccuracies W1 or W2 (cavity tray sloping to the inside leaf or hardly any flashing height at the inside leaf, cavity tray short-circuited by mortar droppings), we have a 100% certainty rain penetration will cause rising damp in the inside leaf (Figure 11.3).

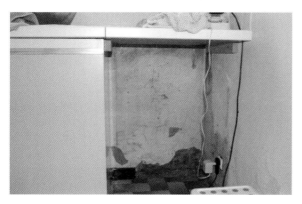

Figure 11.3. Rain penetration: on the left seen as rising damp in the inside, on the right giving a wet inside leaf along window bay.

11.4 Example of risk analysis: cavity walls

Inaccuracy W7 (full fill, tray above the window bays draining sideway on the fill) ends with wet inside leafs along window and outer door bays in 50% of the cases (Figure 11.3). Error D3 (no inside plaster, concrete block inside leaf), finally increases the risk that the inaccuracies W4, W5 and W6 will induce rain penetration.

11.4.5.3 Others

Contrary to mould and rain penetration, inhabitants hardly perceive interstitial condensation as an unwanted consequence. For northwest over south to southeast looking cavity walls, rain buffering by the veneer not only masks the deposit, but the wind vector causes infiltration rather than exfiltration there, which reduces interstitial condensation probability. Northeast, exfiltration dominates but wind driven rain is so to say absent. A veneer wall now may absorb up to 22.5 kg/m^2 of condensation deposit before reaching capillary saturation. At this moisture content, frost damage risk is still close to zero. Only beyond the critical moisture content for frost does damage risk become 1.

With a score of 5 for discontent, the risk interstitial condensation brings seems extremely low. With one out of twenty-five dwellings potentially suffering, we reach a value of 0.2. For bad thermal comfort, risk even does not reach 0.1, whereas a higher energy use for heating means hardly anything, except if the energy prices should increase drastically. In such case, inaccuracy W3 will become the culprit.

11.4.6 Evaluation

Table 11.6. Global risk evaluation.

Error, inaccuracy	Risk				
	Mould	Rain penetration	Interstitial condensation	Thermal comfort	Energy use
D1	1.9				< 0.1
D2	0.95				< 0.05
D3		See D5, D6	< 0.2	< 0.1	0?
D4		See W1			
D5		See D6:			
D6		0.625			
W1		0.3125			
W2		0.125			
W3	17–61				(0.95?)
W4					
W5		See D5, D6			0?
W6					
W7		0.5625			
W8	2.4				
W9					
Sum	22.25–66.25	1.625	< 0.2	< 0.1	< 0.15–(1.1?)

Table 11.6 summarizes the evaluation. In it, the fact that error D1 and D2 and inaccuracy W8 occur in partially and in fully filled cavity walls, while inaccuracy W3 is limited to partial fills, was accounted for. One risk clearly dwarfs all others: mould. In case of full fills evaluated separately, rain penetration also peaks.

11.4.7 Upgrade proposals

Inaccuracy W3 produces the highest risk in terms of malfunctioning: insulation boards not lining up with the inside leaf in case of a partial fill. Related risk is so high that already in the 1980s better execution techniques were proposed:

1. Using outside instead of inside scaffolding. First brick laying and pointing the inside leaf, then insulating, then building the veneer wall
2. Producing specific insulation boards for partial fills with a dense front layer and a sufficiently thick, soft back layer. Fixing with purpose designed screwed ties
3. Developing specific boards for corners and below tray application
4. Introducing prefabricated lintels with everything included (tray, insulation layer, stainless steel section bearing the veneer wall)

Number two in terms of importance are all inaccuracies that favour rain penetration in case of a full fill. Again, the 1980s saw the development of better execution techniques:

1. Again outside scaffolding
2. No full fills thinner than 10 cm
3. Special ties, which exclude sloping to the inside leaf and have in front of the insulation a profile such that rain conduction is excluded.

11.5   References and literature

[11.1]   Hens, H., Fatin, A. M. (1995). *Heat-Air-Moisture Design of Cavity Walls, Theoretical and Experimental Results, Practice*. ASHRAE Transactions 101a.

[11.2]   Sanders, C. (1996). *IEA-Annex 24 Heat, Air and Moisture Transfer in Insulated Envelope Parts*. Final Report, Vol. 2, Environmental Conditions, ACCO, Leuven (1996). 96 pp.

[11.3]   de Wit, M., Schippers, R., Schellen, M. (1997). *The influence of free convection on the thermal resistance of cavity constructions*. Annex 32 paper STB-NL-97/2.

[11.4]   Janssens, A., Hens, H. (1997). *Condensation Risk Assessment*. Internal report Annex 32, STA-B-97/5.

[11.5]   Janssens, A. (1998). *Reliable Control of Interstitial Condensation in Lightweight Roof Systems*. Doctoraal proefschrift, KU Leuven, 217 pp.

[11.6]   Hens, H. (1998). *Envelope performance assessment tools*. Paper for the Copenhagen IEA-Annex 32 meeting, April 1998.

[11.7]   Hendriks, L., Hens, H. (2000). *Building Envelopes in a Holistic Perspective*. Final report IEA Exco ECBCS Annex 41, ACCO, Leuven, 102 pp. + Ad.

11.5 References and literature

[11.8] Hens, H., Carmeliet, J., Roels, S., Janssen, H. (2006). *Whole building approach and hygrothermal risk analysis*. Research in Building Physics and Building Engineering (Eds.: P. Fazio, H. Ge, J. Rao, G. Desmarais), pp. 519–526.

[11.9] Physibel, Trisco Manual (2009).